MAP SKILLS, Grade 6

TABLE OF CONTENTS

INTRODUCTION

The world is constantly changing. Current events carry us to places around the globe. For today's students to be well informed, they must know the basic skills of geography and map use. Map skills help students to improve their sense of location, place, and movement. By giving students a better knowledge of geography and maps, we give them a better understanding of the world in which they live. Such knowledge will help students in their standardized testing and in their broader academic pursuits. And some people need to know how to use a street map just to get across town! *Map Skills* is meant to address these needs and many more.

General standards in geography for this grade suggest that students should be able to perform a variety of skills. Students should be able to understand and use map components, employ the cardinal and intermediate directions, use latitude and longitude, identify the continents and oceans, recognize landforms, and compare various types of information on maps. *Map Skills* provides students with extensive practice in these areas.

Organization

The book is divided into nine units, each centering on a particular map component or type of map. Many of the pages also contain additional activities that put the map skills into practical use. The book also contains a pretest and posttest to assess students' strengths and weaknesses in using maps.

Maps and Globes

Students gain a greater sense of place by knowing their relationship to other places. For this reason, maps and globes are important tools. The book contains a varied group of maps for student use. One or two pages recommend the use of a globe, so if one is available, it should be prominently displayed in the classroom. In any case, various maps, especially of the United States and the world, are a handy addition to any classroom.

Glossary

boundary (p. 13) the dividing line on a map where one place ends and another place begins. A boundary is also sometimes called a **border** (p. 13).

cardinal directions (p. 7) the four main directions of north, south, east, and west

cartographer (p. 36) a person who makes maps

climate map (p. 45) a map that shows the climate in a certain area

compass (p. 8) a device for determining directions

compass rose (p. 7) a symbol that shows the directions on a map

continent (p. 33) a very large body of land. There are seven continents.

cultural map (p. 43) a map that tells about the culture of people in a certain area, including such things as ethnic background, religion, or language

degrees (p. 30) the unit of measurement used for lines of latitude and longitude

direction (p. 7) the line or course along which something is moving or pointing

distance (p. 15) how far one place is from another, often measured in miles or kilometers

distance scale (p. 15) the guide to what the distances on a map stand for

elevation map (p. 24) a map that shows the elevation, or height, of the land's surface. An elevation map is sometimes called a relief map.

Equator (p. 30) the imaginary line that goes around the middle of Earth. The Equator divides Earth into the Northern and Southern Hemispheres.

globe (p. 33) a spherical model of the Earth

grid (p. 11) a pattern of lines that cross each other to form squares or rectangles

hemisphere (p. 35) half of the globe or half of the Earth. The four hemispheres are Northern, Southern, Eastern, and Western.

historical map (p. 39) a map that shows a place during another time in history

inset map (p. 17) a small map within a larger map

intermediate directions (p. 9) the in-between directions of northeast, southeast, southwest, and northwest

intersection (p. 10) the place where two or more routes meet or cross

landform map (p. 23) a map that shows the shape of the land, such as mountains and hills. A landform map is also known as a **physical map** (p. 25) or a relief map.

legend (p. 10) another name for a map key

lines of latitude (p. 30) lines that circle the Earth north and south of the Equator. They are numbered and marked by degrees.

lines of longitude (p. 30) lines that circle the Earth from the North Pole to the South Pole. They are numbered and marked by degrees.

location (p. 7) tells where something can be found

map (p. 7) a drawing of a real place. A map shows the place from above.

map index (p. 12) the alphabetical list of places on a map identified by their grid section

map key (p. 10) the guide to what the pictures or symbols on a map stand for

mileage (p. 18) distance in miles

mileage table (p. 21) a table on a map that tells how far in miles one place is from another

miles per hour (p. 22) a rate of speed, abbreviated as **mph**

national capital (p. 13) a place where laws are made for a nation or country

ocean (p. 33) a very large body of water. There are four oceans.

political map (p. 38) a map that shows the boundaries between political areas, such as states or nations

population map (p. 46) a map that shows the density of population in a certain area

Prime Meridian (p. 30) the line of longitude from the South Pole to the North Pole measured at 0°. It divides the Earth into the Eastern and Western Hemispheres.

projection (p. 36) a way of showing the Earth's curved surface on a flat map

resource (p. 27) something people can use, such as oil, trees, and water

resource map (p. 27) a map that shows the location of resources in a place. A resource map is also sometimes known as a **product map** (p. 27) or a **land use map** (p. 28), which shows how people make use of the land itself.

route (p. 18) a road or path from one place to another. Highways, railroads, and trails are routes.

route map (p. 18) a map that shows the locations and intersections of routes. Route maps are also called **road maps** (p. 19) or **street maps** (p. 19).

sea level (p. 24) the height of 0 feet on an elevation map

state capital (p. 13) the center of government in a state, where laws are made for that state

symbols (p. 10) drawings or patterns that show what things on a map mean

time zone (p. 42) an area of the Earth where the time is the same. The Earth has 24 time zones.

title (p. 10) a name that identifies a map and its contents

vegetation map (p. 44) a map that shows the kinds of vegetation, or plants, that grow in a certain area

Name ______________________ Date ____________

Pretest

Directions Use the map to answer the questions.

THE UNITED STATES

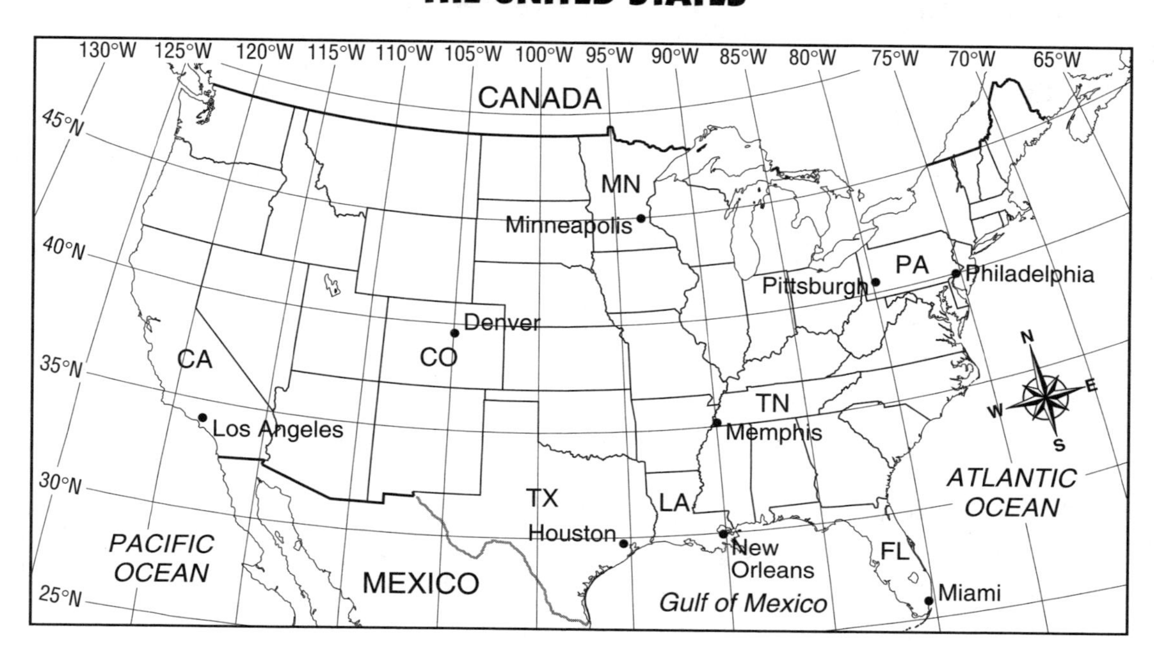

1. What large body of water is located west of Los Angeles?

____

2. What nation does Texas share a border with?

____

3. Which city is at the same latitude as New Orleans?

____

4. Which city is at the same longitude as New Orleans?

____

5. Which three cities are located near 40° N?

____

____

____

6. Which city is located at 26° N, 80° W?

____

7. Which city is located at 45° N, 93° W?

____

Name ______________________ Date ______________

Posttest

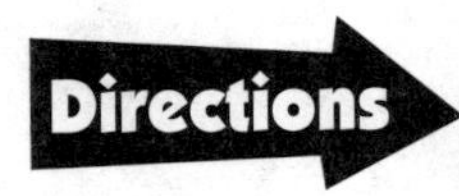

Use the maps to answer the questions. Darken the circle by the correct answer to each question.

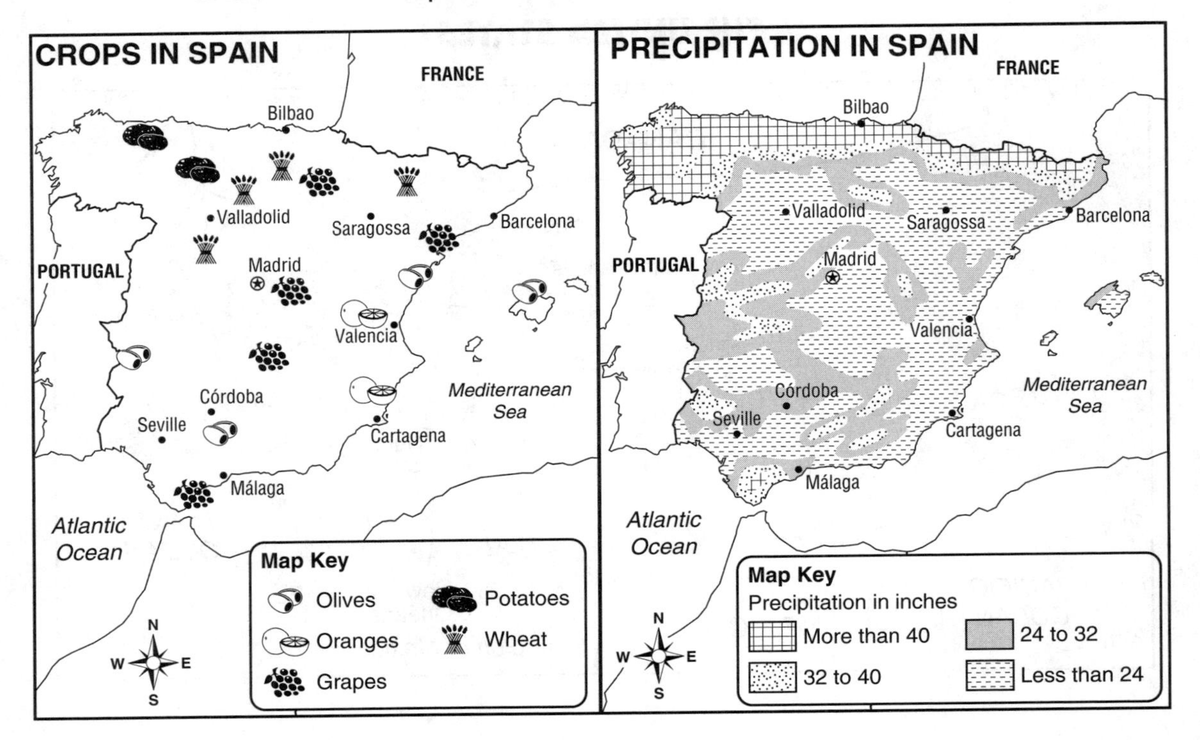

1. What large body of water lies southeast of Spain?

Ⓐ Atlantic Ocean
Ⓑ Gulf of Cartagena
Ⓒ Mediterranean Sea
Ⓓ Map Key

2. What country is located northeast of Spain?

Ⓐ Portugal
Ⓑ France
Ⓒ Crops
Ⓓ Atlantic

3. Which city gets 24 to 32 inches of precipitation and is near an olive-growing region?

Ⓐ Barcelona
Ⓑ Madrid
Ⓒ Valencia
Ⓓ Córdoba

4. Which crop grows in the wettest region of Spain?

Ⓐ oranges
Ⓑ olives
Ⓒ potatoes
Ⓓ peanuts

5. How much precipitation falls in areas where oranges are grown?

Ⓐ more than 40 inches
Ⓑ 32 to 40 inches
Ⓒ 24 to 32 inches
Ⓓ less than 24 inches

6. Barley needs about as much water as wheat does. Near which city might barley be grown?

Ⓐ Valladolid
Ⓑ Seville
Ⓒ Barcelona
Ⓓ Cartagena

Name ______________________________ Date ______________

United States Map

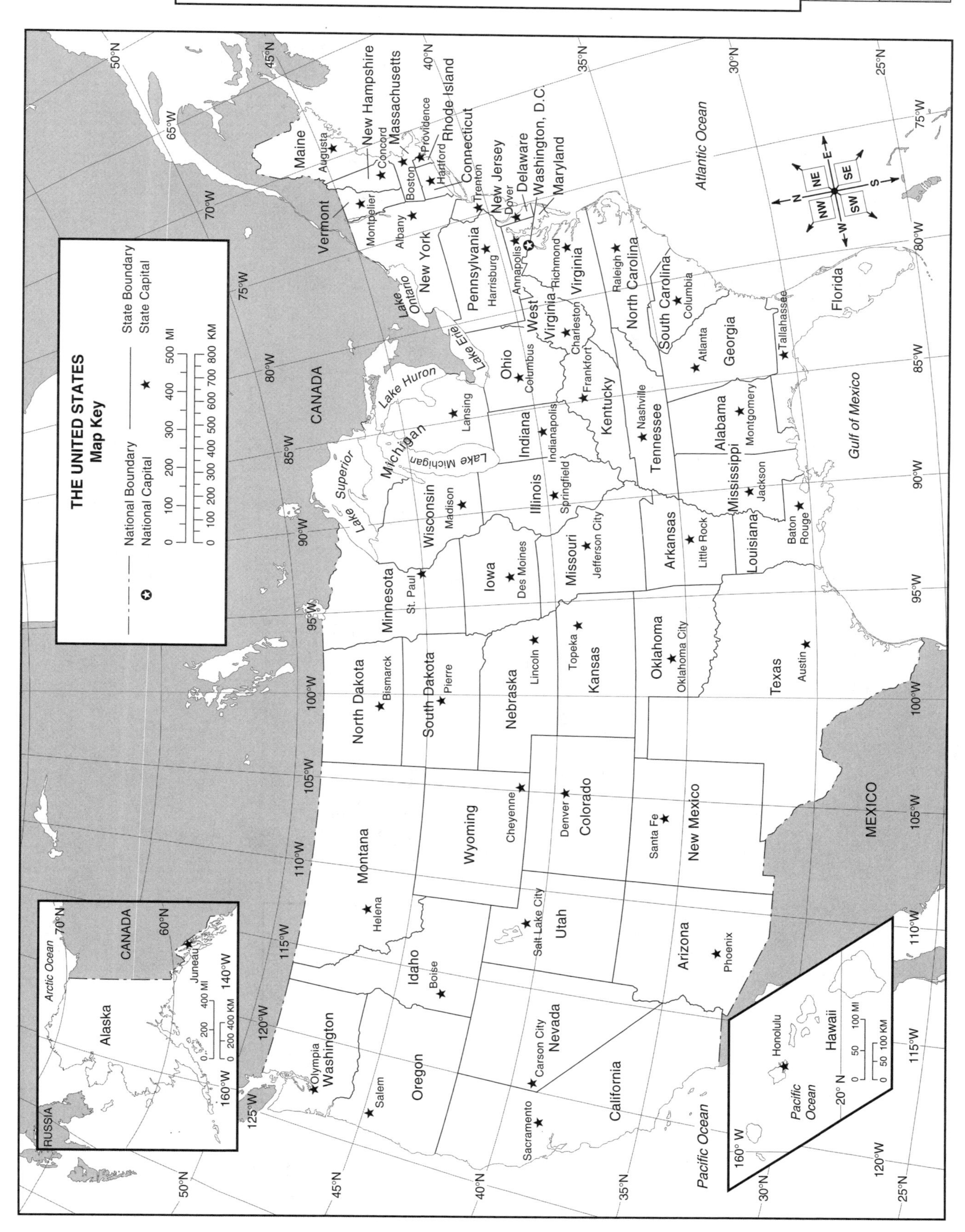

Name ______________________________ Date ______________

World Map

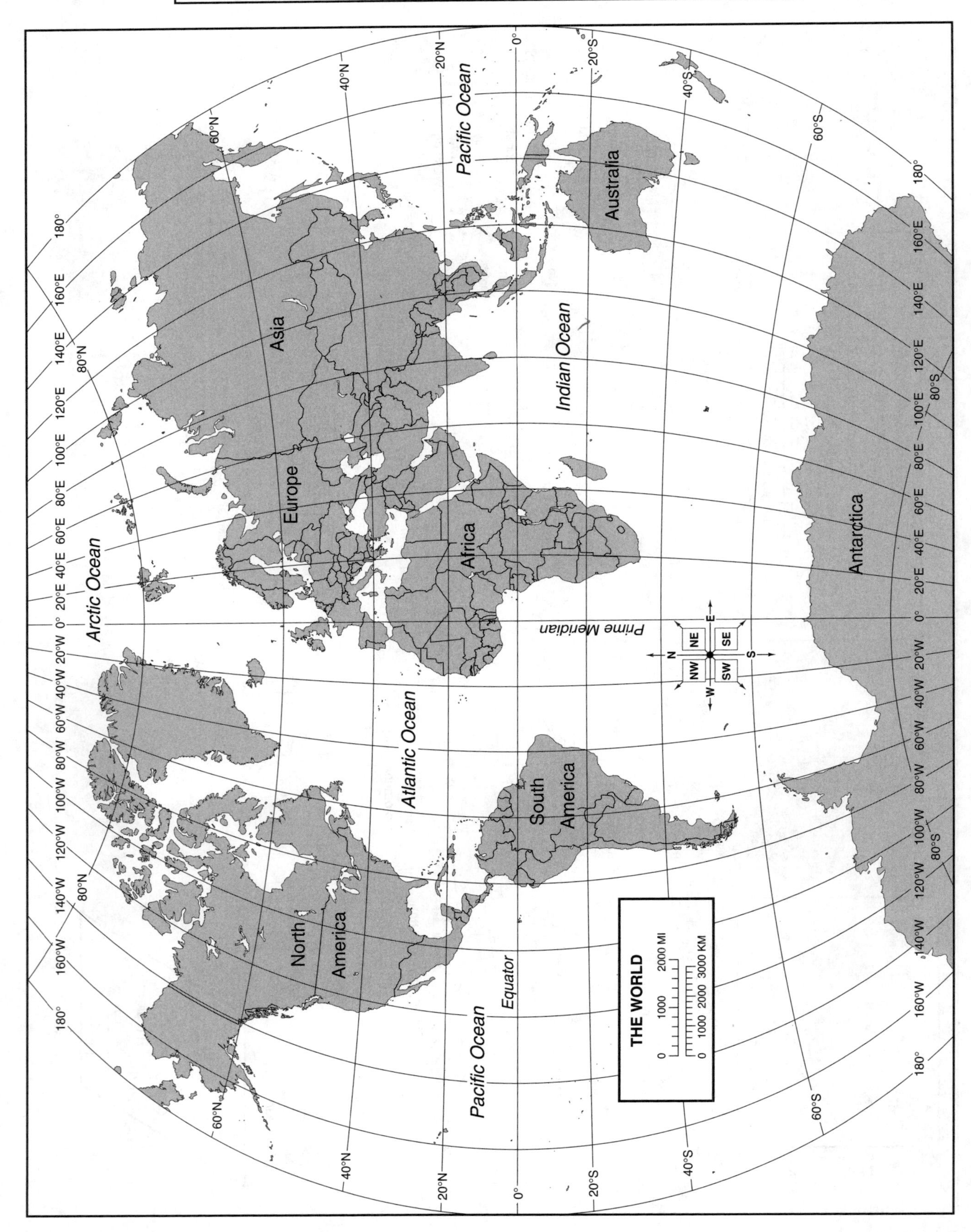

Name ______________________ Date ______________

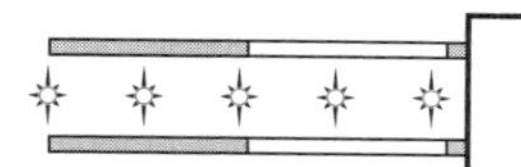

Cardinal Directions

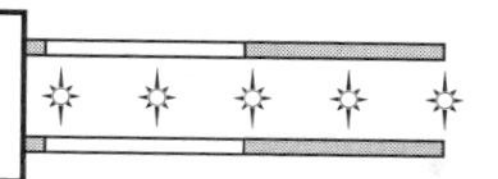

You use **directions** every day. Directions help you to compare **locations**. Directions tell you where you are in relation to other places. They also tell you where other places are in relation to you. You might use directions such as front, back, up, down, right, left, near, or far.

A **map** is a drawing of a real place. It shows the place as if you are looking at it from above. A map uses the directions of north, south, east, and west. These are called the **cardinal directions**. Look on the map for a direction marker. This marker is called a **compass rose**. It may have the abbreviations for all the cardinal directions: *N*, *S*, *E*, *W*. Or it may have only the abbreviation for north: *N*. If it has only the *N*, you have to remember how to find the other directions. Face a wall map or hold the map below in front of you. Find north on the compass rose. It is toward the top of the map. South is the opposite of north, so it is toward the bottom of the map. East is toward the right side of the map. West is toward the left side.

OAK BLUFF

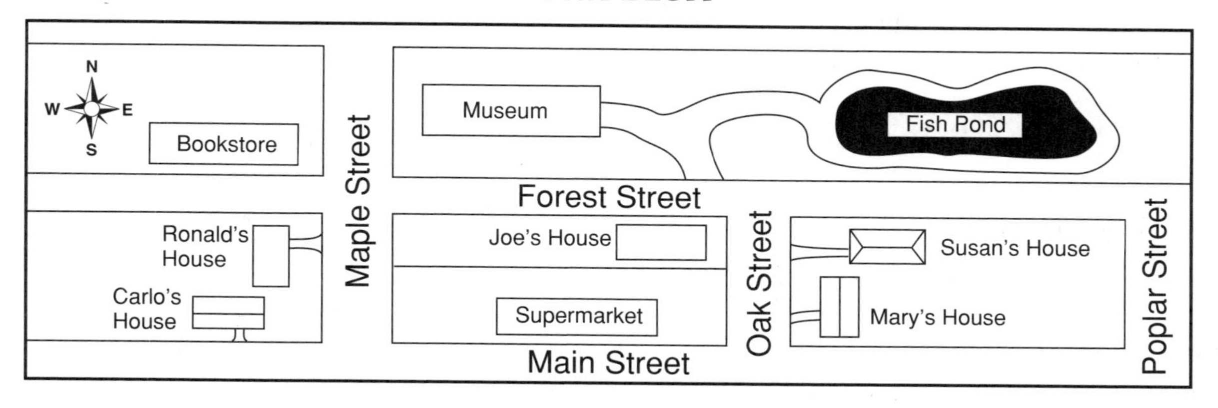

Directions Use the map to answer the questions.

1. Is Mary's house east or west of the supermarket? ______________________

2. Is the bookstore north or south of Carlo's house? ______________________

3. Which directions does Maple Street run? ______________________

4. Which directions does Main Street run? ______________________

Activity

Draw a map of your neighborhood. Include at least four streets. Include the houses on those streets. Put an **X** on your house. Add any other features that you want. When you finish, share your map with the class. Save your map for use in another activity later.

Name ______________________ Date ______________

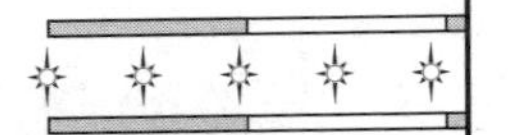

Making a Compass

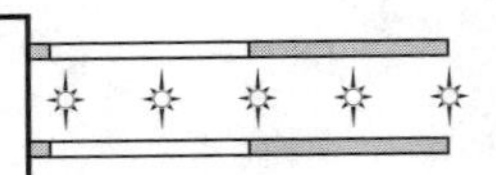

A **compass** is an instrument with a magnetized needle. A compass is used to find the directions of north, south, east, and west. The needle of the compass always points north. If you know where north is, you can find all the other directions. You can make your own compass. Here's how.

You will need:

1 sewing needle (It's sharp, so be careful.)
1 magnet (such as one on the refrigerator)
1 small plastic bowl of water
1 small piece of paper, about 2 inches square
1 real compass

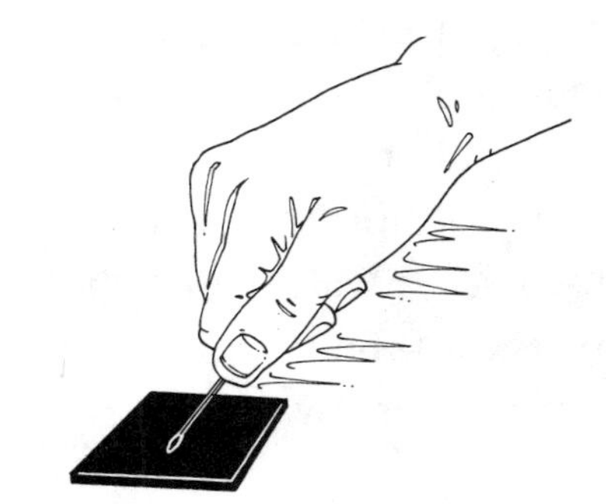

Steps:

1. Stroke the **dull** end of the needle across the magnet. Do this about 60 times. Be sure to stroke the same direction each time.
2. Put the bowl of water on a table. The bowl should not be made of metal. Make sure there is no metal within 2 feet of the bowl. The table should not be metal. Check to see there is no metal under the table. Also, make sure the magnet you used is not nearby. The real compass should not be close, either.
3. Now, set the paper on top of the water in the center of the bowl. Set the needle on top of the paper. Move the paper to make it spin slightly. If the paper gets stuck on the side of the bowl, move it back to the center.
4. Watch the needle. What is happening? The needle and paper should stop moving completely. When they do, the **sharp** end of the needle should be pointing north.

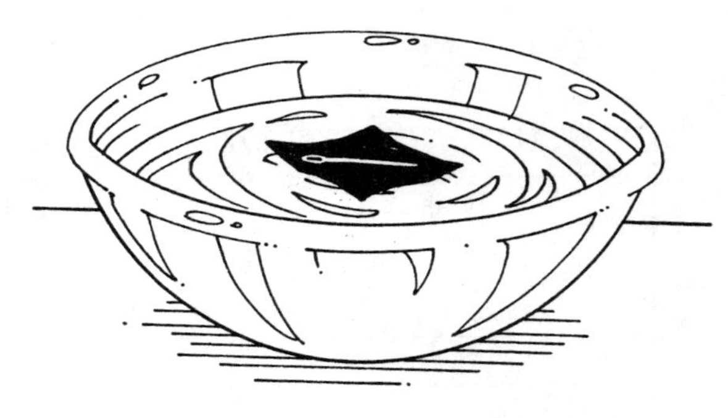

5. Check how accurate your homemade compass is. Use the real compass. Don't get the two compasses too close to each other. If you do, they will interfere with one another.
6. Now, you know which direction is north. Face north. South will be behind you. East will be to your right. West will be to your left.

Activity

Imagine that you are lost in the woods. The time is 4 P.M. The Sun is still shining brightly, but it is sinking in the sky. You are worried that you might not get out of the woods before dark. You know that you have to go south to get back to the campground. It would be easy to find your way out with a compass, but you forgot the compass in your tent. How will you get out of the woods? Write a paragraph telling what you would do.

Name ______________________ Date ______________

Intermediate Directions

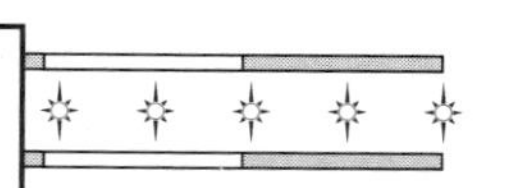

The compass rose on a map shows the four main directions. These are the cardinal directions: north, south, east, west. Some maps also include in-between directions. These are the **intermediate directions:** northeast, southeast, southwest, northwest. They fall between the cardinal directions. The intermediate directions are abbreviated *NE, SE, SW, NW.*

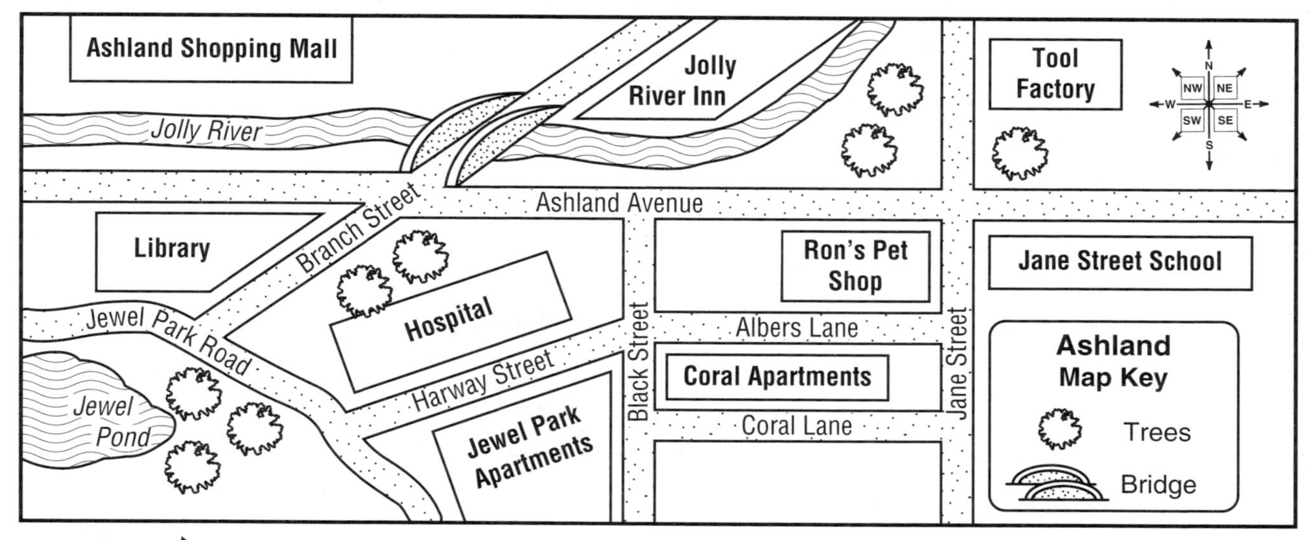

Directions Study the map of Ashland. Then, darken the circle by the correct answer to each question.

1. The hospital is ______ of the Jewel Park Apartments.
- Ⓐ northwest
- Ⓑ southwest
- Ⓒ southeast
- Ⓓ northeast

2. The ______ is northeast of Ron's Pet Shop.
- Ⓐ Jolly River Inn
- Ⓑ Coral Apartments
- Ⓒ Jane Street School
- Ⓓ Tool Factory

3. Teri lives in the Coral Apartments. She attends the Jane Street School. Which direction must she go to get to school?
- Ⓐ northwest
- Ⓑ southwest
- Ⓒ southeast
- Ⓓ northeast

4. Branch Street runs ______.
- Ⓐ northwest and southeast
- Ⓑ northwest and northeast
- Ⓒ northeast and southwest
- Ⓓ northwest and southwest

Activity

The compass rose on a map helps you to find directions. Sometimes, though, it is hard to use. You can make a compass rose that moves. It will be easier to use. Get a small piece of clear plastic about 2 inches square. On it, draw your own compass rose with a black marker. Put the four cardinal directions on the four points. Add the intermediate directions, too. Now, you can put the compass rose anywhere on the map. Try it out on the map above. Remember, north points toward the top of the map.

Name ______________________________ Date ______________

Map Parts

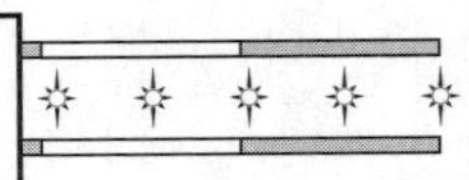

To use a map well, you must know about its parts. These parts help you to know what information the map contains. You already know about the compass rose. The compass rose indicates the directions. A map usually has a **title**. The title tells what the map is about. Most maps also have a **map key**, or **legend**. Maps use **symbols**, which are drawings or patterns to show what things on the map mean. The map key tells you what these symbols mean.

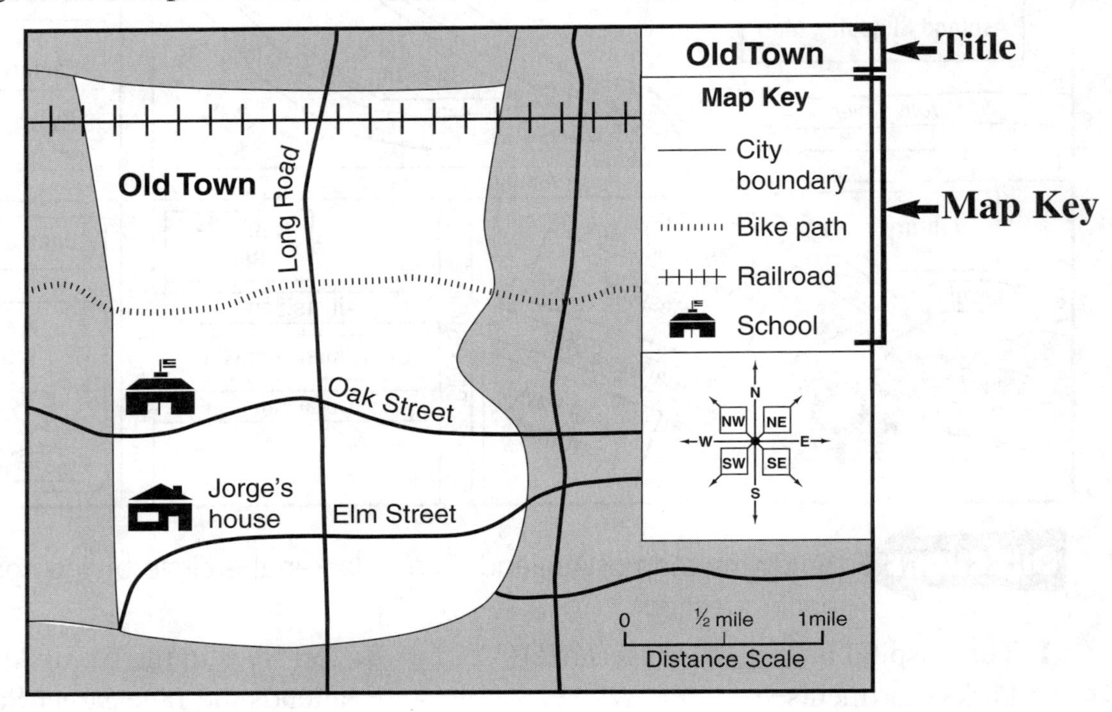

Directions Look at the map of Old Town. Find the title and map key. Study these parts carefully. Then, answer the questions.

1. What is the title of this map? ______________________________

2. What does this symbol [school symbol] in the map key stand for? ______________________________

3. What does this symbol [bike path symbol] in the map key stand for? ______________________________

4. Find Jorge's house on the map. Jorge must walk to the **intersection** of Oak Street and Long Road. Which direction should he go? ______________________________

Activity

Use your neighborhood map from page 7. Add the title "My Neighborhood." Also add a map key that uses symbols for the houses, trees, and other features that you have on your map.

Name ______________________ Date ______________

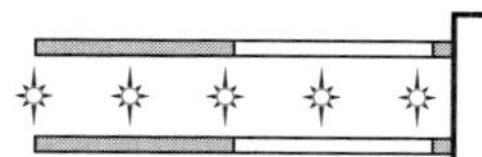

Using a Map Grid

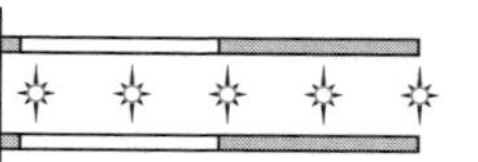

Some maps have a grid to help you to locate places. A map **grid** is a pattern of lines that cross each other. The lines form squares or rectangles. Each square or rectangle is a grid section. Each row of squares or rectangles has a letter. Each column of squares or rectangles has a number.

Look at the map of the Caribbean Islands. The letters are on the left and right sides. The numbers are on the top and bottom. Each rectangle on the map is named with a letter and a number. Find grid section B-2 on the map. Put your finger on the letter **B**. Move your finger along the row. Stop when you get to column **2**. What is in grid section B-2? Jamaica and parts of other islands are in that section.

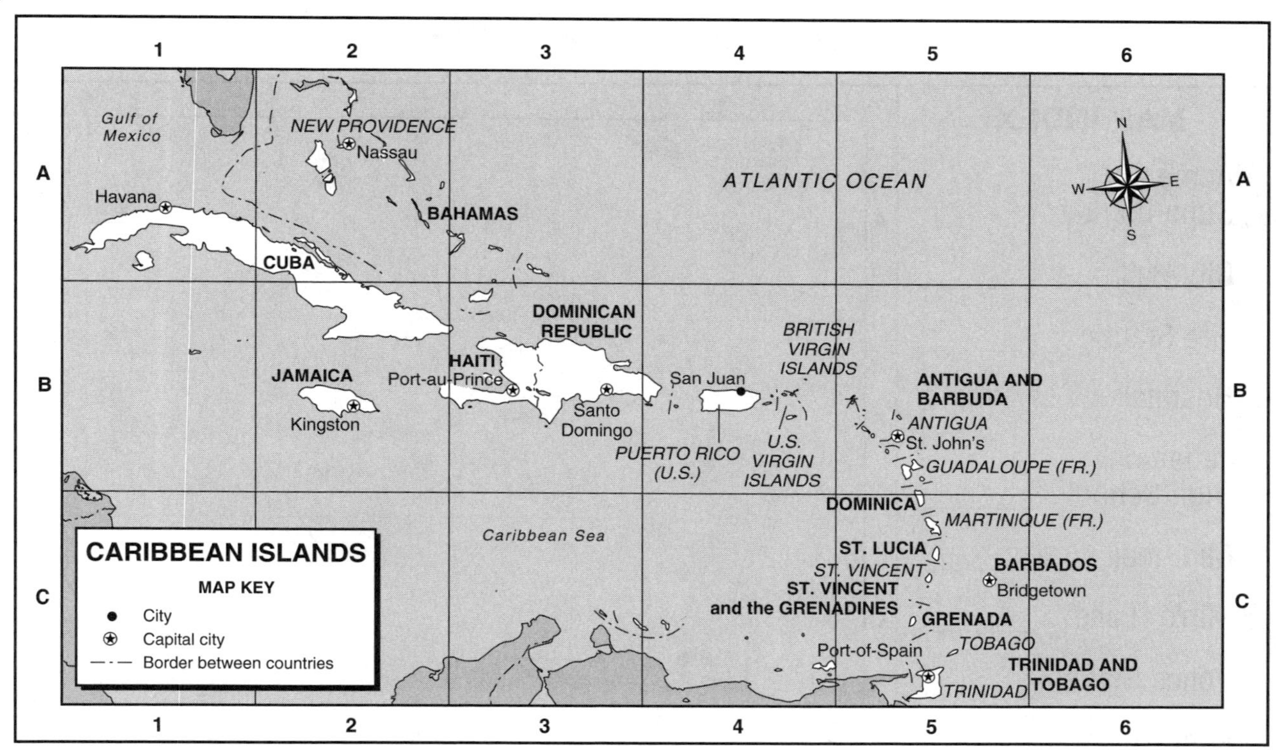

Directions Use the map to answer the questions.

1. In which grid section is Puerto Rico? ______________
2. In which grid section is Grenada? ______________
3. Grid section A-4 is in what body of water? ______________
4. In which grid section are Haiti and the Dominican Republic mainly located? ______________
5. What island is in grid sections A-1, A-2, and B-2? ______________

Activity

Use your neighborhood map from page 10. Draw a grid on your map. Put letters on the left and right sides. Put numbers at the top and bottom. In which grid section is your house?

Name ______________________ Date ____________

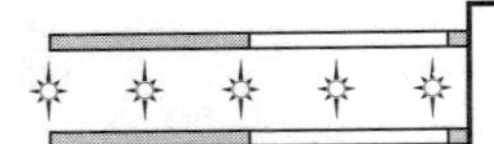

Making a Map Index

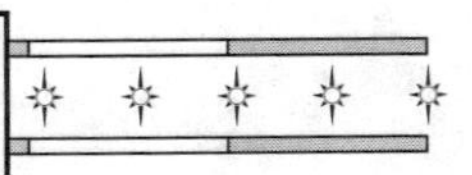

A **map index** is an alphabetical list of places on a map. The map index tells you in which grid section a place is located. Suppose you need to find the hospital. Look at the words that start with **H** in the map index below. Read the grid section letter and number by the word Hospital. The hospital is in grid section C-3. Now, find the hospital on the map.

Directions Complete the map index by finding the grid section for each place.

CITY MAP OF SMITHVILLE

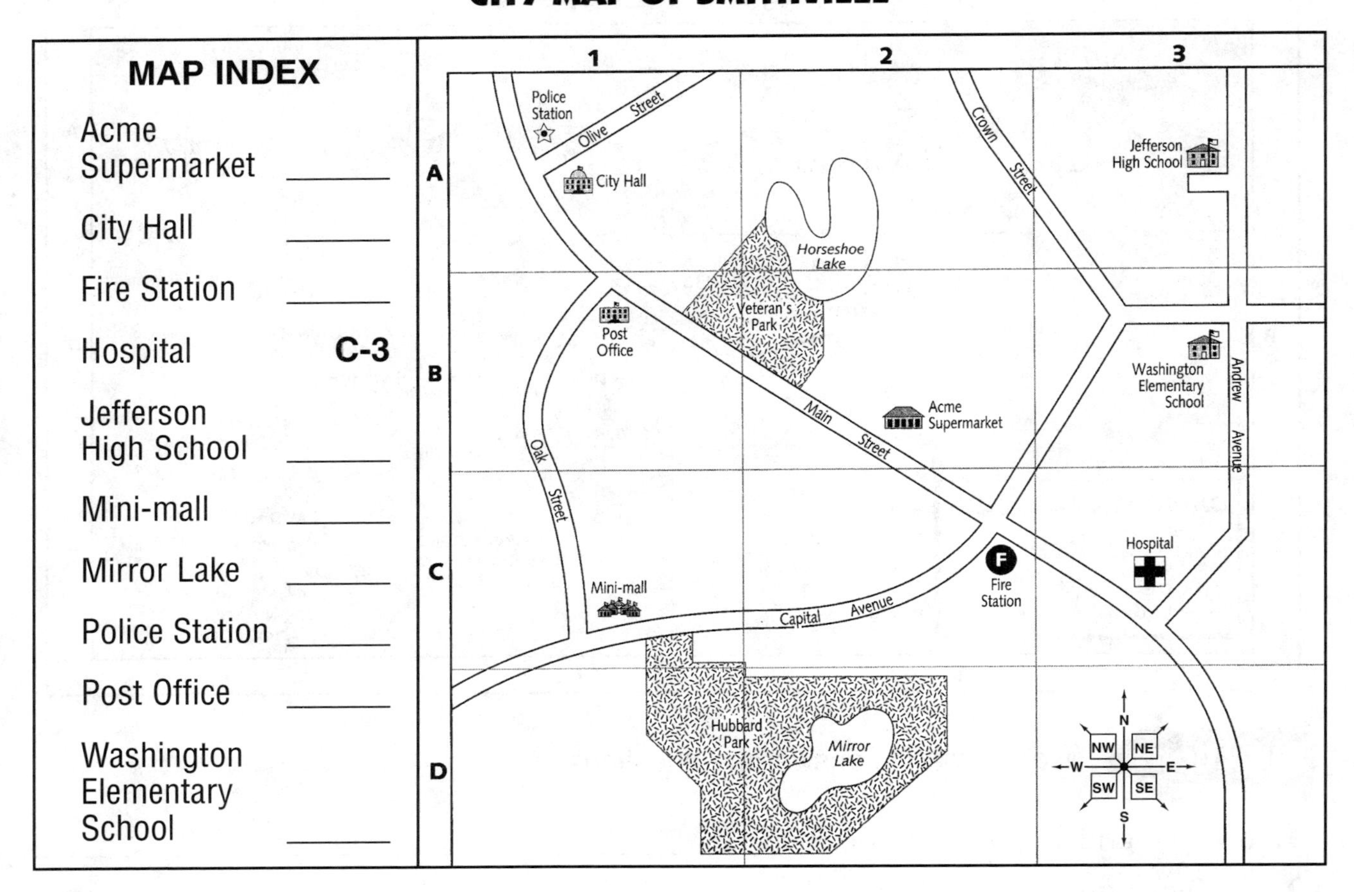

MAP INDEX	
Acme Supermarket	______
City Hall	______
Fire Station	______
Hospital	**C-3**
Jefferson High School	______
Mini-mall	______
Mirror Lake	______
Police Station	______
Post Office	______
Washington Elementary School	______

Activity

Use your neighborhood map from page 11. Use the grid to make a map index. Include at least six terms. One term should be "My House." Put the terms in alphabetical order. Write the grid square letter and number by each term.

Name ______________________________ Date ______________

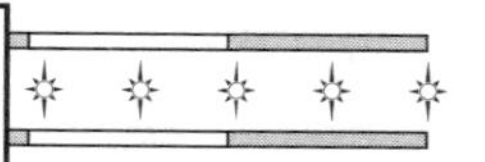

Capitals and Boundaries

The United States is divided into 50 states. Each state has a center of government. That center of government is called the **state capital**. Laws for the whole state are made in the state capital. Do you know the capital of your state? Look at the map key below. On this map, a star stands for a state capital. Can you find the state capital of Washington?

Nations also have capitals. Laws for the whole nation are made in the **national capital**. On many maps, a star in a circle stands for a national capital.

This map also shows state and national **boundaries**. Boundaries are the lines on a map that show where one state or nation ends and another state or nation begins. The boundaries separate the two places. Boundaries are also called **borders**. Find the boundary symbols in the map key. Move your finger along the boundaries on the map.

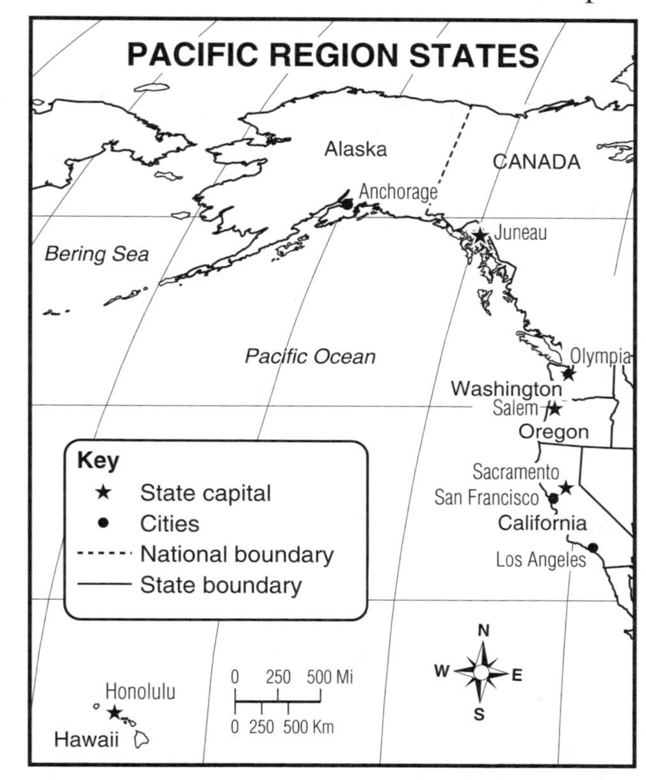

Use the map to answer the questions.

1. What is the state capital of California?

____

2. Look at the map key. A dotted line is a national boundary. With what nation does Alaska share a border?

____

3. If you wanted to travel from the state capital of Oregon to Anchorage, Alaska, which direction would you go?

____

4. If you wanted to travel from the state capital of Alaska to the state capital of Hawaii, which direction would you go?

____

5. What is the state capital of Alaska?

____

6. What is the state capital of Hawaii?

____

7. What is the state capital of Oregon?

____

Activity

A state's shape is determined by its border. Some states have unusual shapes. Colorado is a rectangle. The lower part of Michigan looks like a mitten. Several states have "panhandles." Find your state on the United States map on page 5. What shape does your state have? Move your finger along the state border.

Name ______________________ Date ______________

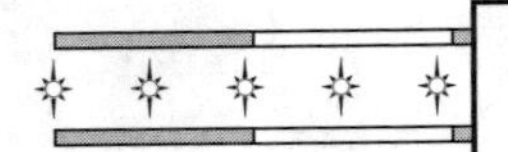

Mapping Your Progress

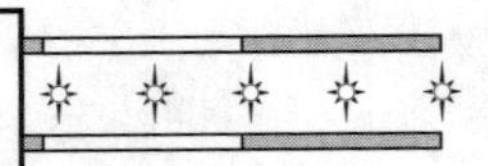

Now you know many of the things necessary to use a map. Let's see how good your map skills are so far.

Study the map of the United States on page 5. Use the map to answer the questions.

1. What is the title of this map?

____

2. Find the symbol for a national capital in the map key. What is the national capital of the United States?

____

3. What is the state capital of New Hampshire?

____

4. What is the state capital of New Mexico?

____

5. What nation shares a border with North Dakota?

____

6. What nation shares a border with Arizona?

____

7. You want to travel from the state capital of South Dakota to the state capital of North Carolina. Which direction do you go?

____

8. You want to travel from the state capital of Connecticut to the state capital of Arkansas. Which direction do you go?

____

9. What is the state capital of the state directly east of Oregon?

____

10. What is the state capital of the state directly south of Georgia?

____

Name ______________________________ Date ______________

Using a Distance Scale

Maps are not life-size. They show big areas on small pieces of paper. **Distance** on a map is not the same as the real distance in a place. To show distance, maps use a **distance scale**. A distance scale shows that a certain length on the map equals a certain length in a real place. Most distance scales show distance in miles and kilometers. Look at the map below. Find the distance scale. On this distance scale, 1 inch = 300 miles. Notice that this map also shows distances in kilometers.

Suppose you want to find the distance from Cairo to the Mediterranean Sea. One way is to use an inch ruler. Measure the distance between the two places. The distance is about $\frac{3}{8}$ inch. On the distance scale, $\frac{3}{8}$ inch = about 100 miles.

EGYPT

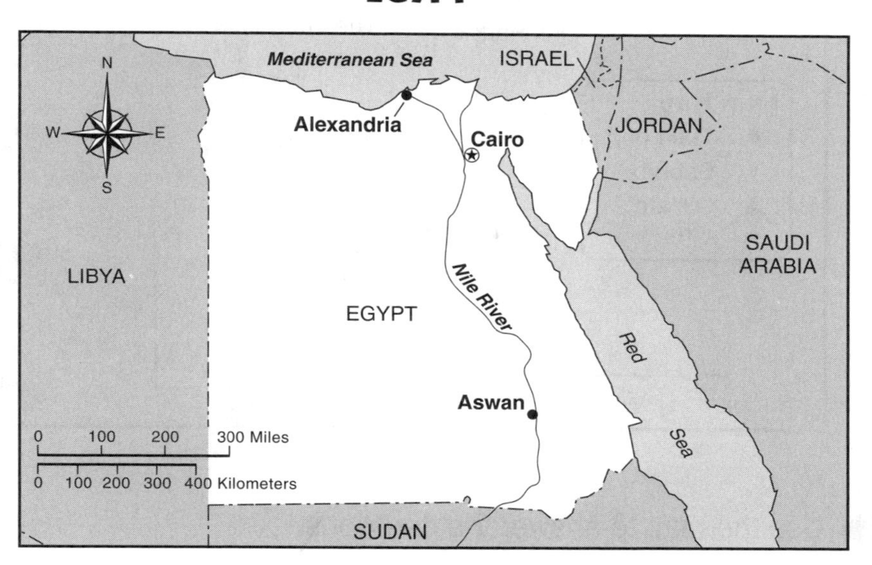

Directions Use an inch ruler and the distance scale on the map to finish each sentence.

1. On the map, the distance from Aswan to Cairo is about ____________ inches.

2. On the map, 1 inch = 300 miles. In the real place, Aswan is about ____________ miles from Cairo.

3. On the map, 1 inch = 480 kilometers. In the real place, Aswan is about ____________ kilometers from Cairo.

4. On the map, the distance from Alexandria to Cairo is about ____________ inch.

5. In the real place, Alexandria is about ____________ miles from Cairo.

6. Asyut is a city about 250 miles northwest of Aswan. Asyut is about 200 miles south of Cairo on the Nile River. Place the city of Asyut on the map. Draw a dot and write the name Asyut.

Name ______________________ Date ______________

How Far Is It?

On a city or area map, the distances between places are smaller. On a city map, 1 inch might equal only a mile or two. Look at the map of the Chicago area below. On this map, 1 inch equals about 8 miles.

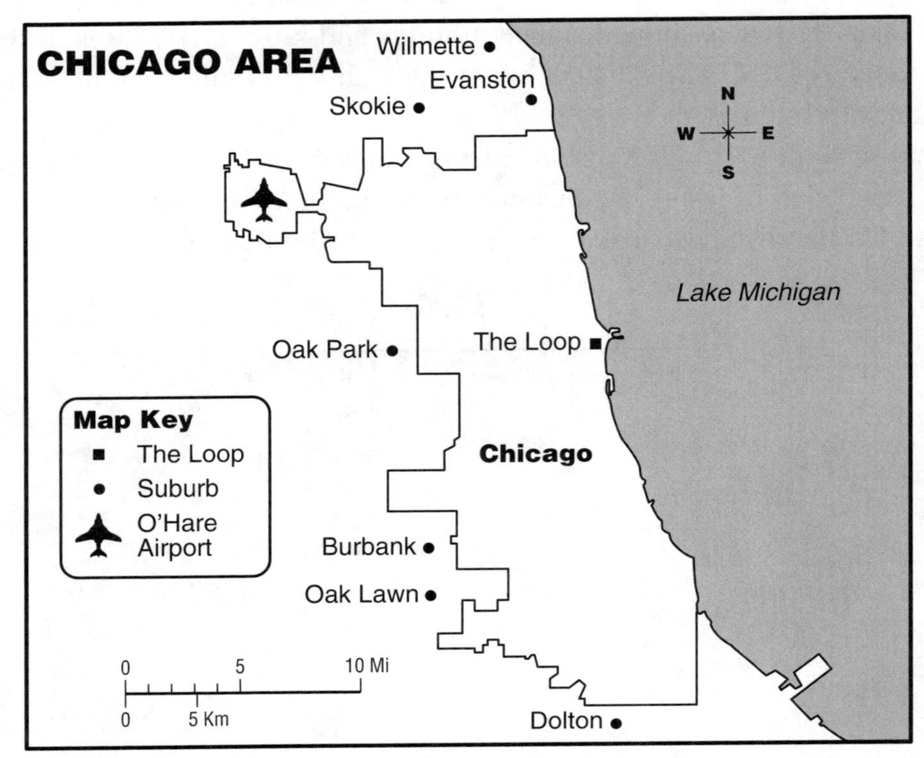

Directions Use the map to answer the questions.

1. About how many miles is Oak Park from the Loop? ______________

2. About how many kilometers is Dolton from Oak Lawn? ______________

3. About how many miles is Skokie from Burbank? ______________

4. About how many miles is Dolton from Skokie? ______________

5. About how many kilometers is Wilmette from O'Hare Airport? ______________

6. About how many miles is the Loop from Evanston? ______________

7. About how many miles is the Loop from O'Hare Airport? ______________

8. About how many kilometers is Wilmette from Evanston? ______________

Name ______________________ Date ____________

Using Maps with Different Scales

Sometimes you may look at two maps that show the same general area. You may have one map that gives more detailed information about a smaller area. You may have another smaller map, or **inset map**, that shows where the detailed area is in relation to a larger area. Look at the maps below. The smaller map in the bottom right corner shows Spain, Portugal, and part of France. The larger map shows a detailed view of the southern part of Spain. Notice the distance scale for each map. On the larger map, 1 inch equals about 90 miles. On the smaller map, 1 inch equals about 350 miles.

SPAIN AND ANDALUSIA

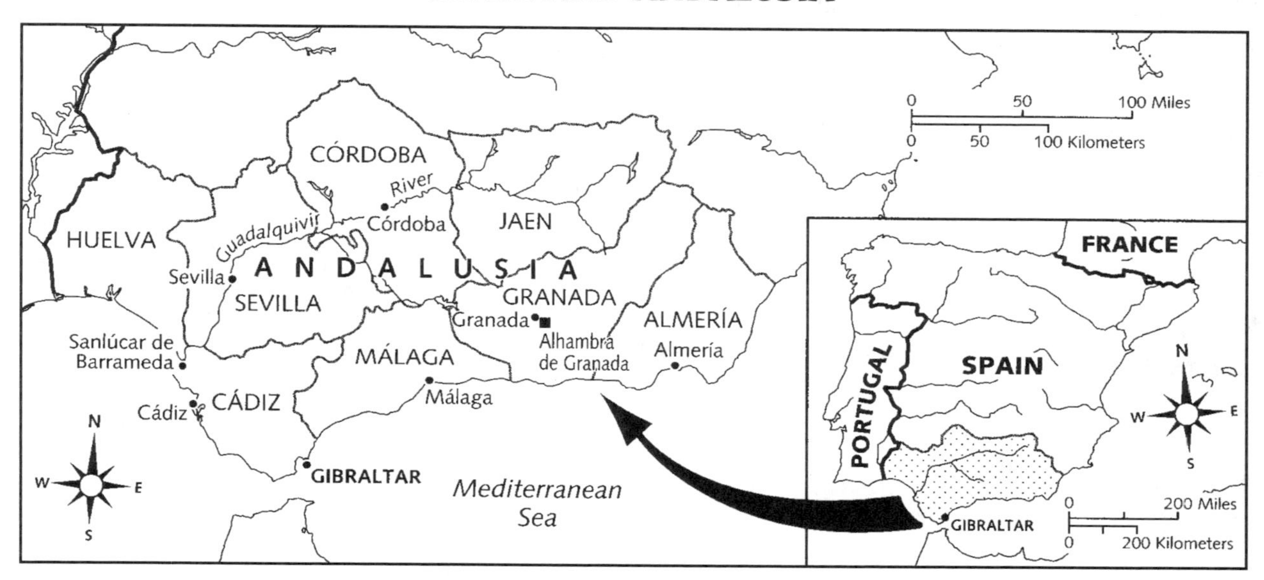

Directions Study the two maps. Use the maps to answer the questions.

1. Which map is better for measuring the distance between Gibraltar and the city of Sevilla?

   ____

2. About how many miles is Gibraltar from the city of Sevilla? ____________

3. Which map is better for measuring the distance between Gibraltar and the border with France? ______________________

4. About how many kilometers is Gibraltar from the French border? ____________

5. You need to find the Alhambra de Granada, which was once a military stronghold of the Moors. Which map would you use to find it? ____________

Name ______________________ Date ______________

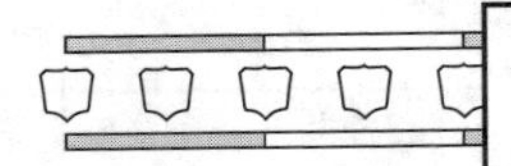

Route Maps

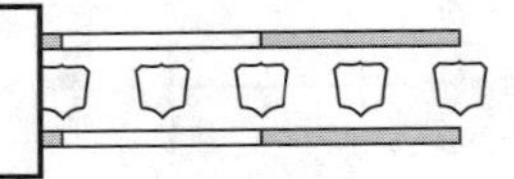

A **route** is a way to get from one place to another. When people go on trips, they often use maps to help them find their way. A good **route map** makes the travel easier. Many route maps show highways. There are several kinds of highways. There are interstate highways, United States (or U.S.) highways, and state highways. Each kind of highway has a different kind of sign. Study the map key. It shows the different kinds of highways.

Some route maps do not use a distance scale. Instead, they have numbers to show distance in miles, or **mileage**. Look at the Jones County map. See the **9** between Collier and Jerzy? That number means the distance between the two places is 9 miles. Let's say you need to drive from Jerzy to Collier and then on to Bullock. How far is the trip? Add the mileage. The trip is 17 miles.

JONES COUNTY

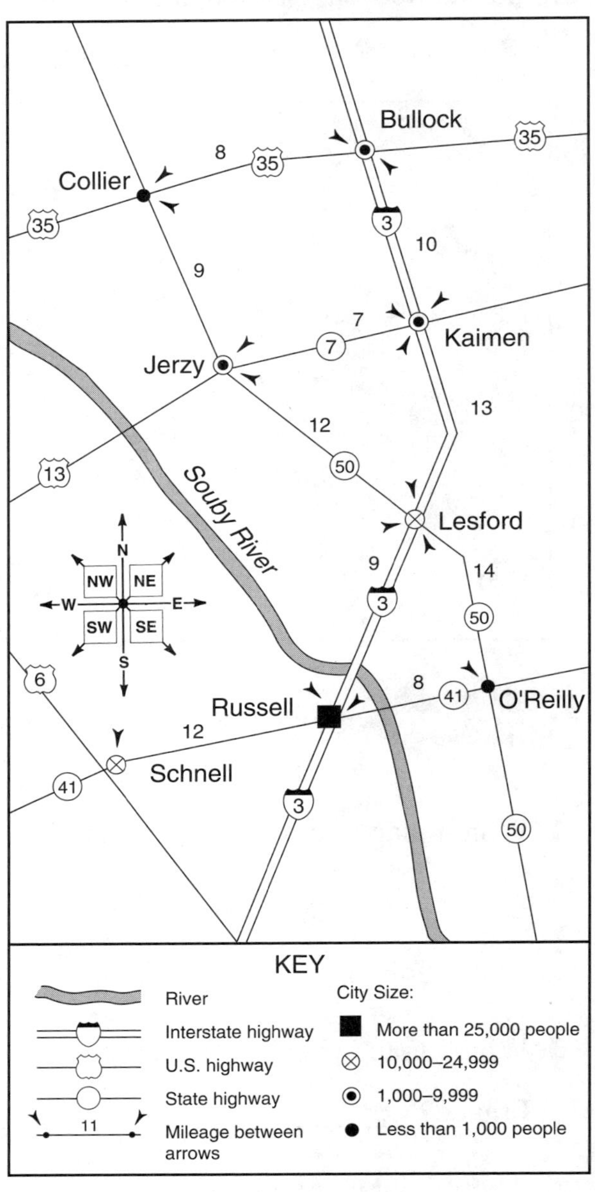

Directions Study the map. Use the map to answer the questions.

1. What highway runs from Jerzy to Kaimen?

____

2. What is the shortest distance from Russell to Collier?

____

3. What would be the shortest route from O'Reilly to Bullock?

____

____

4. In which general directions does State Highway 50 run?

____

Activity

Kara and her dad live in O'Reilly. Kara's dad wants to go on an afternoon drive. He wants to drive about 80 miles round-trip. Plan a route for the afternoon drive. Include highway numbers, directions, and distances.

Name ______________________________ Date ______________

Let's Go to Town

Sometimes route maps are called **street maps** or **road maps**. Many people use street maps to get around their city. The street maps show all the streets in the city. Most street maps have a distance scale, too. A person can see how to go from one place to another and how far apart they are. Sometimes, the person must make many turns and go on many streets to travel somewhere.

ATLANTA, GEORGIA

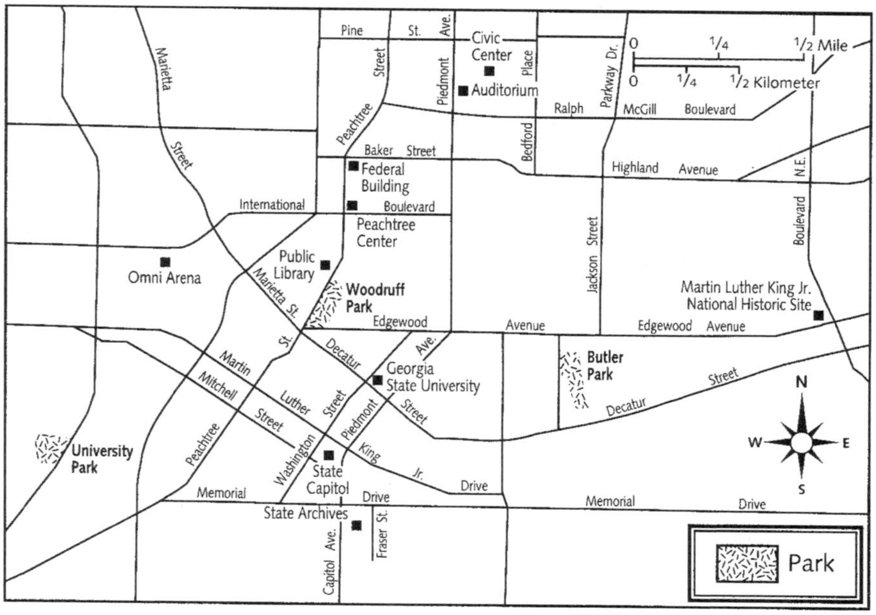

Directions Study the map of Atlanta. Use the map to answer the questions.

1. You are at the State Capitol. You need to go to the Peachtree Center. What route would you use? Include street names and directions.

____

____

2. About how far would you have to go on your route from the State Capitol to the Peachtree Center?

____

3. You are at Omni Arena. You want to go to the Martin Luther King, Jr., National Historic Site. What route would you use? Include street names and directions.

____

____

4. Start at the intersection of Piedmont Avenue and Pine Street. Go west on Pine Street. Go south on Peachtree Street. Go east on Edgewood Avenue. Go southwest on Washington Street. Go east on Memorial Drive. Go south on Capitol Avenue. Where are you?

____

Activity

Find a route from the State Capitol to some other place on the map. Then, write the set of directions. Give the directions to a classmate. Can the classmate find the place?

Name ______________________ Date ____________

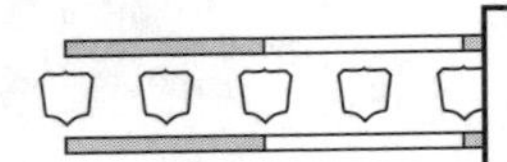

Historical Routes

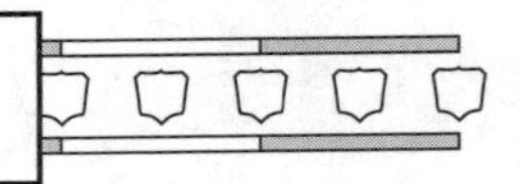

Routes have changed over time. In the 1800s, people used trails to head west. They found many adventures along the Oregon Trail or the California Trail. By the 1900s, highways had replaced the trails. One of the most famous of these highways was Highway 66. Find this highway on the map, and follow its route with your finger.

In the 1930s, the middle section of the United States had a drought. The land dried out and turned to dust, causing the area to become known as the Dust Bowl. Many people left the Dust Bowl and headed west, looking for a better life.

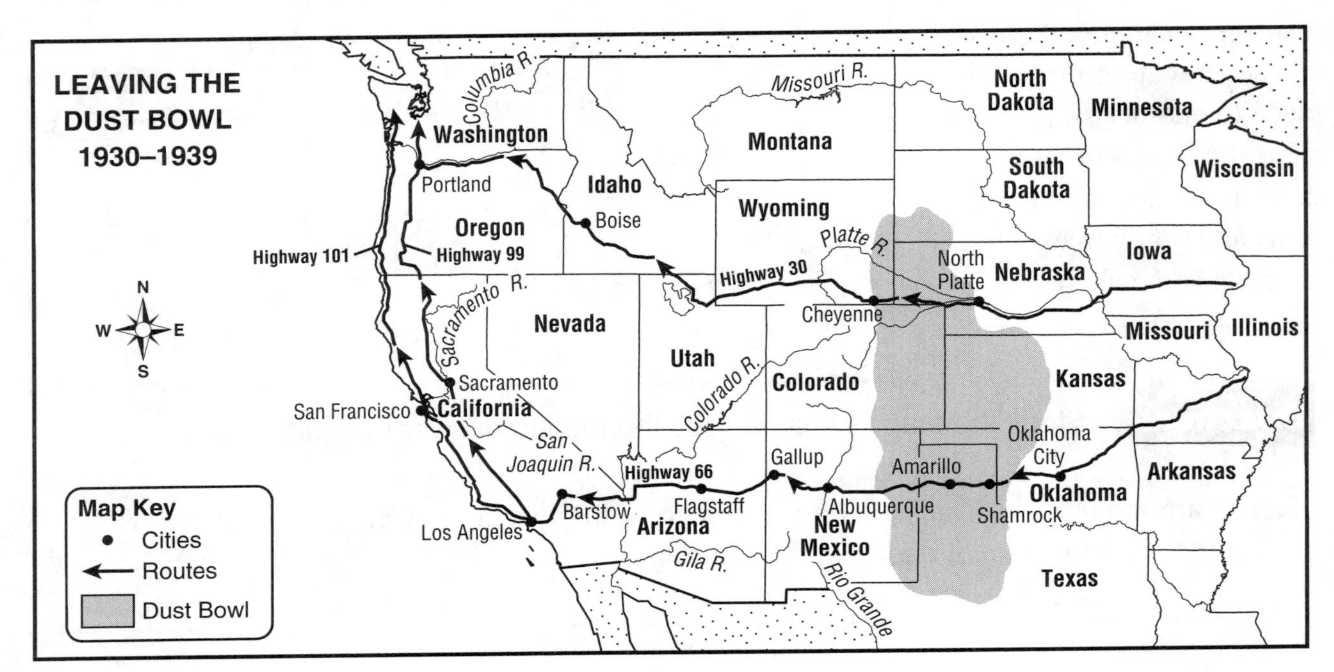

Directions Use the map to answer the questions. Darken the circle by your answer choice.

1. Which highway crosses the Rio Grande?
- Ⓐ Highway 101
- Ⓑ Highway 66
- Ⓒ Highway 30
- Ⓓ Highway 99

2. Which highway runs along the western coastline of the United States?
- Ⓐ Highway 101
- Ⓑ Highway 66
- Ⓒ Highway 30
- Ⓓ Highway 99

3. Which highway runs from Iowa to Washington?
- Ⓐ Highway 101
- Ⓑ Highway 66
- Ⓒ Highway 30
- Ⓓ Highway 99

4. Which of these cities is not located on Highway 66?
- Ⓐ Gallup
- Ⓑ Amarillo
- Ⓒ Barstow
- Ⓓ Boise

Activity

Imagine that you and your family were forced to leave your home and travel to a distant place. How would you feel? What would you do? Write a story about what might happen to you.

Name ______________________________ Date ________________

Using a Mileage Table

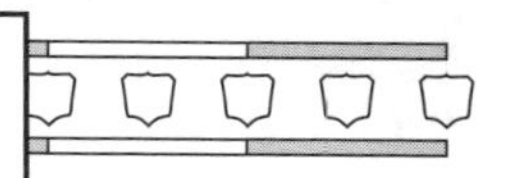

Some route maps include a **mileage table**. A mileage table tells you how far one place is from another place. Study the mileage table below. Five city names are listed on the left side. Those same names are at the top. Let's say you want to know how many miles Glendive is from Great Falls. Find Glendive on the side. Run your finger to the right to the Great Falls column. The distance is 353 miles. You can also find Glendive on the top. Run your finger down the column to the Great Falls row. The distance is 353 miles.

MONTANA: MAJOR ROADS

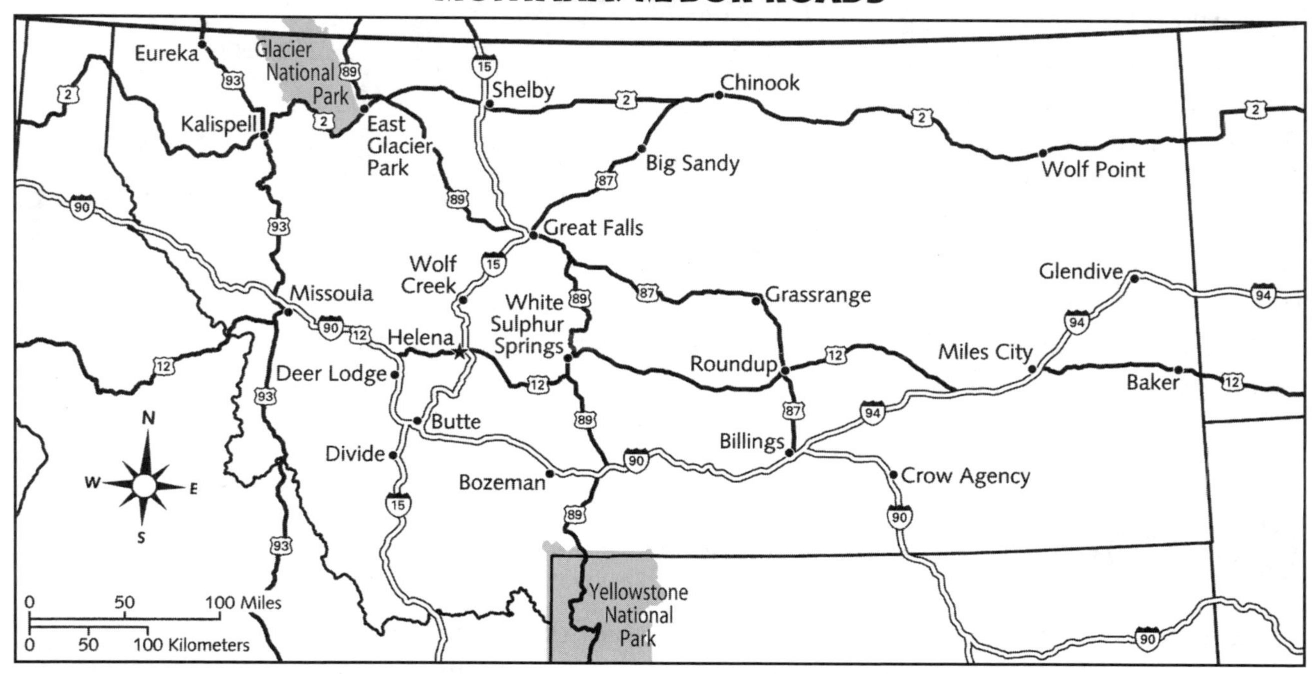

	Billings	Butte	Glendive	Great Falls	Kalispell
Billings		236	221	224	452
Butte	236		457	157	240
Glendive	221	457		353	581
Great Falls	224	157	353		228
Kalispell	452	240	581	228	

Study the map and the mileage table. Use the mileage table to determine the distance for each of these trips. Then, trace each trip on the route map using a different color.

1. Billings to Glendive ________________

2. Kalispell to Butte ________________

3. Butte to Glendive ________________

4. Glendive to Kalispell ________________

5. Kalispell to Great Falls ________________

6. Great Falls to Billings ________________

Name ______________________________ Date ______________

Figuring Travel Time and Distance

To plan a trip well, you need to know how far you will travel. You also need a good estimate of how long the trip will take. You can find how far the trip will be by using the distance scale. Measure the distance between the places so you have a good idea of how many miles you must travel. Then, figure the travel time. Many times, drivers on a highway average 60 **miles per hour (mph)**. On a street in a city, an average speed might be 30 miles per hour. And if you are walking, an average speed might be 4 miles per hour. To figure the travel time, divide the distance by miles per hour. For example, if the distance is 180 miles and you travel at 60 miles per hour, the travel time will be 3 hours.

SOUTH DAKOTA HIGHWAYS

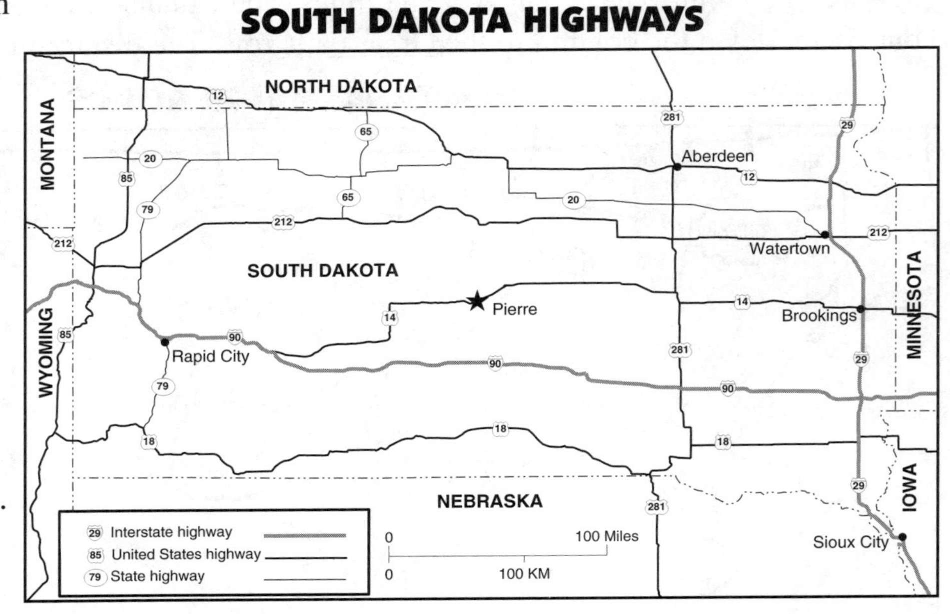

Directions Use the map to answer the questions.

1. You must travel from Pierre to Brookings. You plan to travel 60 miles per hour. How long will the trip take?

____

2. You plan a drive from Rapid City to Aberdeen. What route will you take?

____

____

3. On your drive from Rapid City to Aberdeen, you will travel 50 miles per hour. How long will the drive take?

____

4. You must drive from Watertown to Sioux City, Iowa. How long will the drive be?

____

5. How long will the trip take if you average 55 mph?

____

6. You leave Pierre at 10 A.M. to drive to Rapid City. If you average 40 mph, when will you arrive in Rapid City?

____

Activity

Find a road map of your state. Find your city. Plan a trip to another city that will take 2 hours if you average 55 mph.

Name ______________________ Date ______________

Landform Maps

A **landform map** tells about the form, or shape, of the Earth's surface. It gives information on various physical features, such as mountains, hills, plateaus, plains, and deserts. It can also tell where rivers are. Maps indicate these natural features in different ways. Some maps use drawings to show the landforms. Other maps use shading. The map below has the names of the landforms printed on the map.

MAJOR LANDFORMS AND NATURAL REGIONS OF THE SOUTHWEST

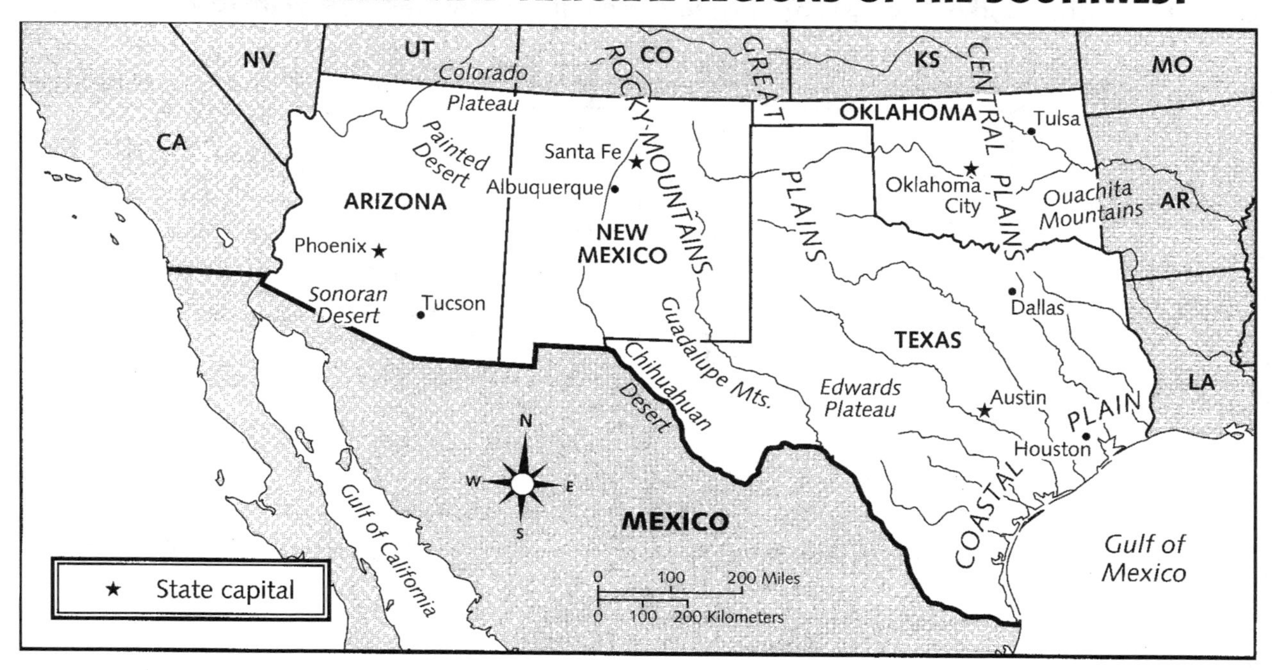

Directions Use the map to answer the questions.

1. Which landform covers the southwestern part of Arizona? ______________
2. In which natural region is Oklahoma City located? ______________
3. Which city is shown at the edge of the Rocky Mountains? ______________
4. What mountain range is located in southeast Oklahoma? ______________
5. In which natural region is Houston located? ______________
6. Which three deserts are shown on the map? ______________

   ____
7. Which two plateaus are shown on the map? ______________

 **Activity**

What are some landforms in your state? Do research to find out. Then, use the map on page 5 to draw an outline of your state. Write the names of the landforms where they are located in your state.

Name ______________________ Date ______________

Elevation Maps

A landform map is sometimes called an **elevation map**. An elevation map shows the elevation, or height, of the land's surface. Study the elevation scale on the map below. It shows the elevation in feet and meters. The scale begins at 0 feet, which is also called **sea level**.

The map below uses scattered dots to show elevation from 0 to 250 feet. Notice these dots along the coastal plain of Texas. A crisscross pattern is used to show elevation from 2,000 to 5,000 feet. Much of West Texas shows this pattern. Elevations above 5,000 feet are shown with dark vertical lines. Other patterns are used to show the other elevation ranges.

TEXAS ELEVATION MAP

Directions Study the map and the elevation scale. Use the map to answer the questions.

1. The highest parts of Texas are located in what part of the state?

____

2. The lowest parts of Texas are located in what part of the state?

____

3. Is Lubbock at a higher or lower elevation than Corpus Christi is?

____

4. What is the elevation range of the area around Amarillo?

____

5. Look back at the map on page 23. What is the name of the physical feature around the Houston area?

____

6. What is the elevation range of this physical feature?

____

Name ______________________________ Date ______________

Comparing Information on Maps

Sometimes landform maps are called **physical maps**. They show the physical features of a place, such as mountains, plains, and rivers. If you study a physical map, you can often learn why human features, such as cities or routes, are built where they are.

ITALY: PHYSICAL

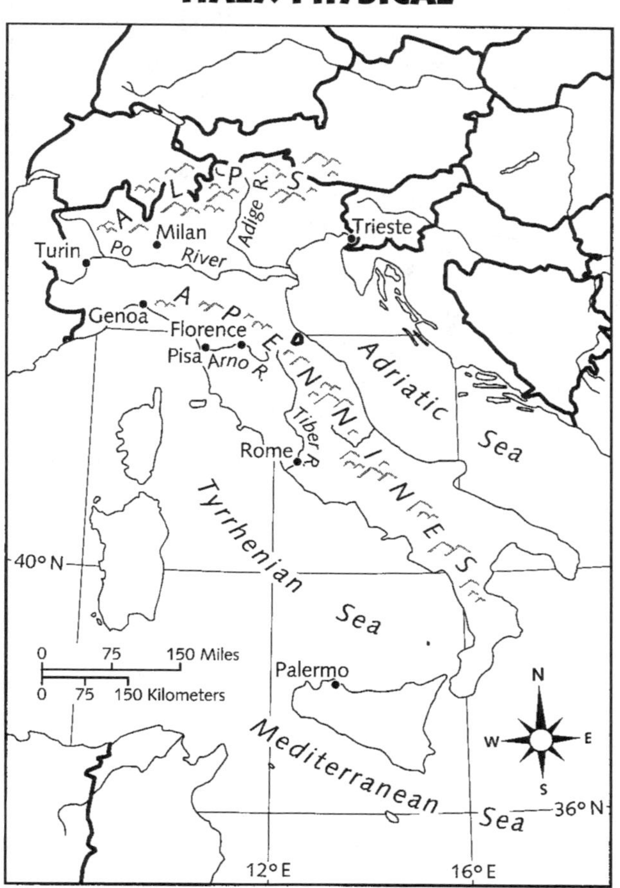

ITALY: MAJOR ROADS

Directions Study the two maps. Use the maps to answer the questions.

1. What kind of information does each map provide? ______________________________

____

2. What is the relationship between Italy's cities and mountains? ______________________

____

3. What is the relationship between the mountains and the location of the roads? __________

____

Name ______________________________ Date ______________

Rivers

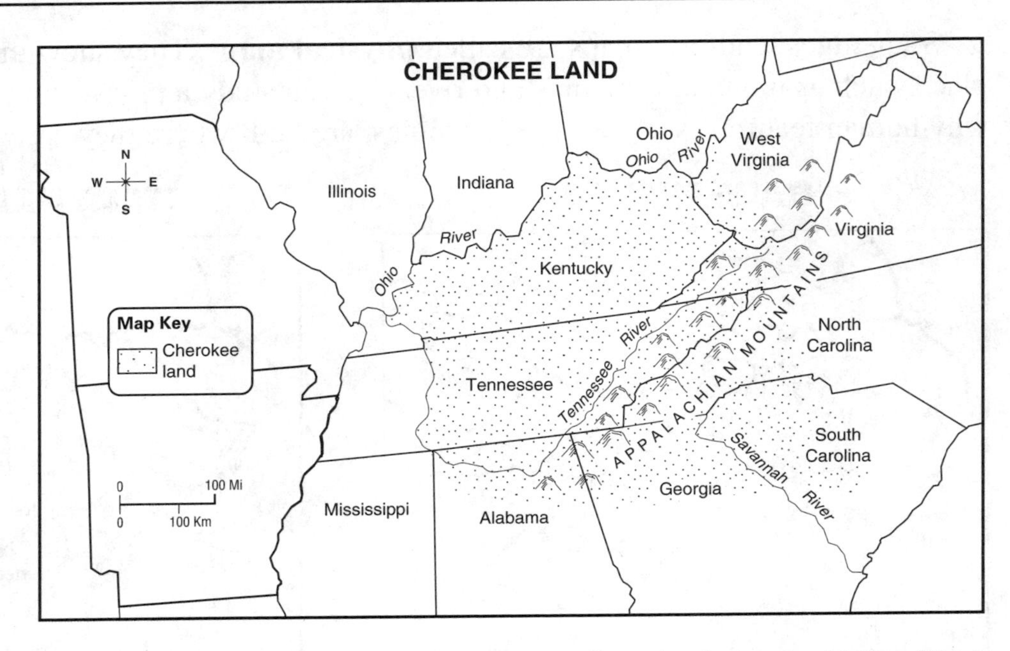

One important physical feature is a river. There are many major rivers in the United States. Rivers are useful for moving goods and people. Rivers are also useful for recreation, supplying drinking water, or generating power. Some rivers serve as boundaries between states or nations.

Study the map. It shows the land lived on by the Cherokee Indians in the early 1800s. Find the rivers. Notice the other landforms, too.

Use the map to answer the questions. Darken the circle by the correct answer to each question.

1. Which river serves as the northern boundary of the Cherokee land?
- Ⓐ Ohio River
- Ⓑ Tennessee River
- Ⓒ Savannah River
- Ⓓ Mississippi River

2. Which mountain range was part of the Cherokee land?
- Ⓐ Rocky Mountains
- Ⓑ Tennessee Mountains
- Ⓒ Appalachian Mountains
- Ⓓ Virginia Mountains

3. Which river runs along the western edge of the Appalachian Mountains?
- Ⓐ Ohio River
- Ⓑ Tennessee River
- Ⓒ Savannah River
- Ⓓ Alabama River

4. Which river serves as a border between South Carolina and Georgia?
- Ⓐ Ohio River
- Ⓑ Tennessee River
- Ⓒ Savannah River
- Ⓓ Virginia River

Activity

Mark Twain wrote a famous novel about Huck Finn's travels by raft down the Mississippi River in the mid-1800s. Imagine that you are Huck Finn. What do you think you would see on your journey? On another piece of paper, write a story about your adventures. Include a drawing of you and your raft.

Name ______________________ Date ______________

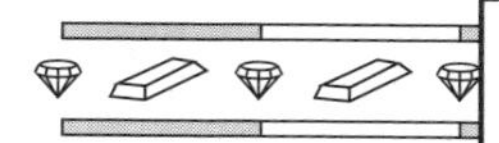

Resource and Product Maps

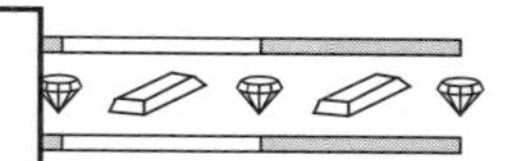

A **resource** is something that people use to make or produce things. Trees are a resource that people use to make lumber and paper. Oil is another resource that people use to make gasoline and plastics. There are many kinds of resources. A **resource map** shows where these resources are located. Sometimes a resource map is called a **product map**. To use a resource or product map, first read the title of the map. It tells you what the map is about. Then, study the map key. The map key will have symbols that stand for resources or products. In the map below, a picture of a tree stands for wood products.

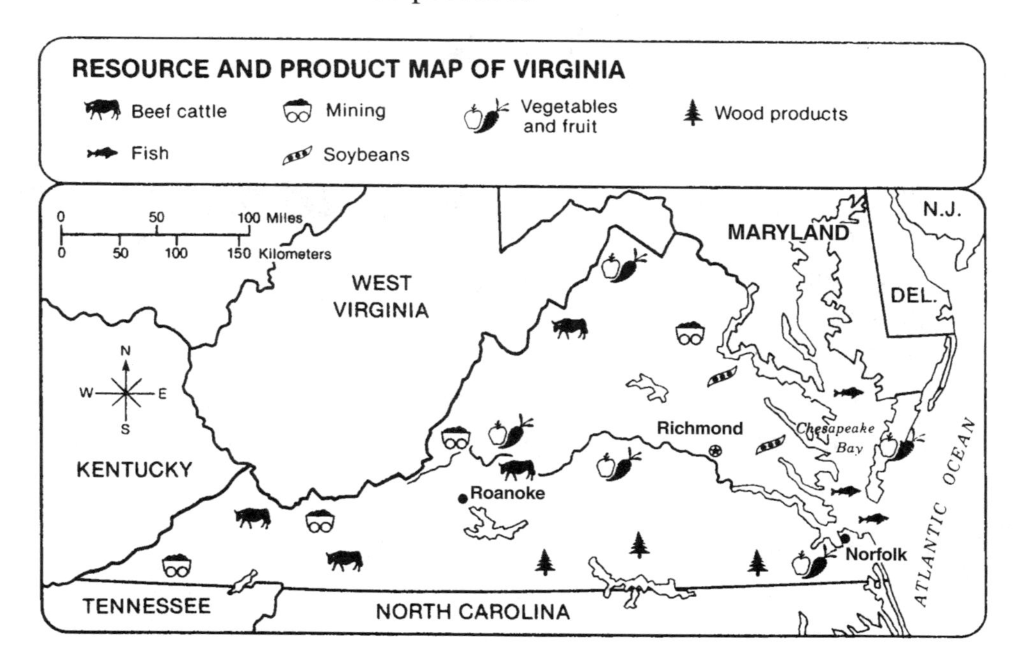

Directions Study the map and map key. Use the map to finish each sentence. Darken the circle by your answer choice.

1. The product found north and east of Richmond is ______.
- Ⓐ cattle
- Ⓑ wood products
- Ⓒ soybeans
- Ⓓ vegetables and fruit

2. The main resources near Virginia's border with West Virginia are ______.
- Ⓐ cattle and mining
- Ⓑ fish and cattle
- Ⓒ soybeans and wood products
- Ⓓ wood products and cattle

3. An important resource southeast of Roanoke is ______.
- Ⓐ cattle
- Ⓑ wood products
- Ⓒ soybeans
- Ⓓ vegetables and fruit

4. The main product in the Chesapeake Bay area of eastern Virginia is ______.
- Ⓐ cattle
- Ⓑ wood products
- Ⓒ fish
- Ⓓ chickens

Name ______________________________ Date ______________

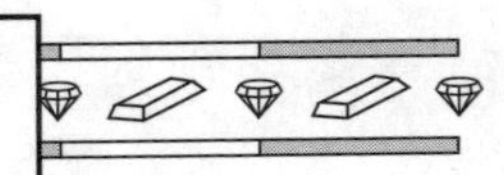

Land Use Maps

Sometimes, resource maps are known as **land use maps**. Land use maps tell how people use the land to produce things. The land can be used for farming, for ranching, for lumber production, for manufacturing, and many other ways. Some maps show resources, products, and land use. These maps allow you to see all the uses people find for their land. Study the map of the imaginary state of East Albion. Notice all the ways people make use of the land.

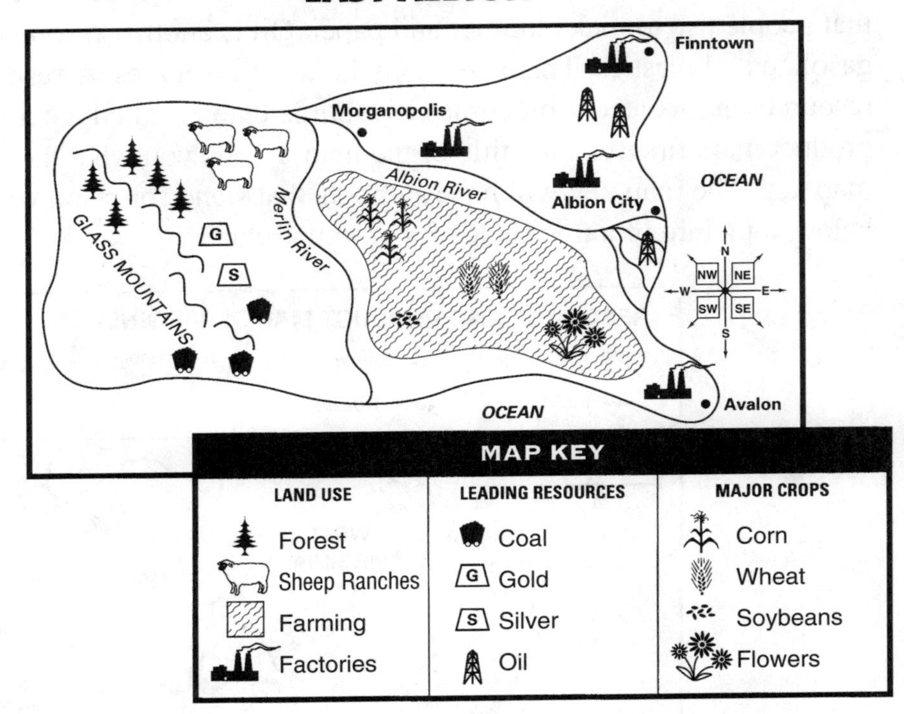

Directions Look closely at the symbols in the map key. Use the map to answer the questions. Darken the circle by the correct answer to each question.

1. Which part of East Albion is used mainly for manufacturing?

Ⓐ northwest
Ⓑ northeast
Ⓒ southwest
Ⓓ central

2. How is the land between the Merlin River and the Albion River mainly used?

Ⓐ farming
Ⓑ manufacturing
Ⓒ oil production
Ⓓ mining

3. Which part of East Albion is used more for forests?

Ⓐ northeast
Ⓑ southeast
Ⓒ southwest
Ⓓ northwest

4. Which three resources are found near the Glass Mountains?

Ⓐ gold, silver, and oil
Ⓑ coal, gold, and oil
Ⓒ coal, gold, and silver
Ⓓ coal, silver, and oil

Activity

What are some resources in your state? Do research to find out about resources in your state. Then, use the map on page 5 to help you to draw a map of your state. Make a map key for your map. Draw pictures for the resources you have learned about. Draw the pictures on the map to show where the resources are found in your state.

Name ______________________ Date ____________

Using a Product Map to Make Generalizations

A product map shows how the land is used to produce different things. Some maps show not only products, but also land use and physical features. By studying the physical features, you can better guess how people might use the land. Would people have farms on the sides of mountains or in the desert? To use such a map, first read the title of the map. It tells you what the map is about. Then, study the map key. The map key has symbols that stand for different things on the map. In the map to the right, different patterns show land use and physical features. Products for each area are printed on the map.

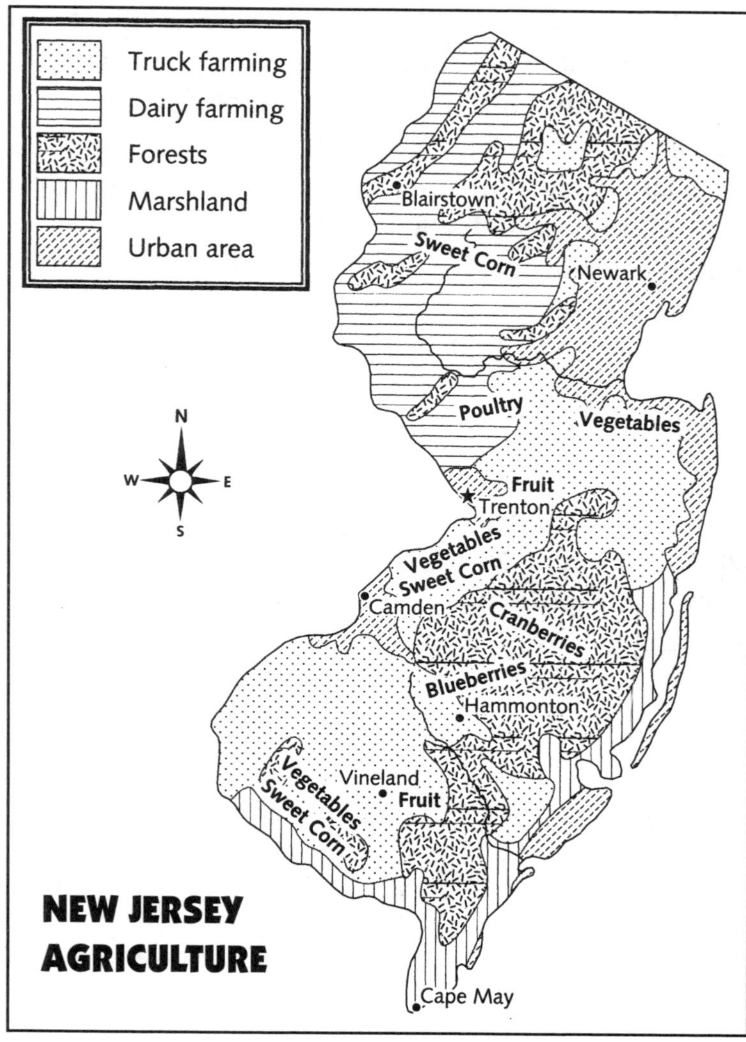

Use the map to answer the questions.

1. Where is the largest urban area in New Jersey?

____

2. Which city is located in the area where cranberries and blueberries are grown?

____

3. What is the main use of the land around Vineland?

____

4. What two crops are produced in the forests of central New Jersey?

____

____

5. Would you be more likely to be a farmer if you lived in Vineland, Cape May, or Newark? Explain.

____

____

____

Name ______________________ Date ____________

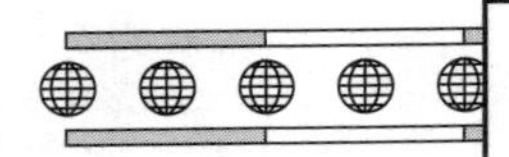

Latitude and Longitude

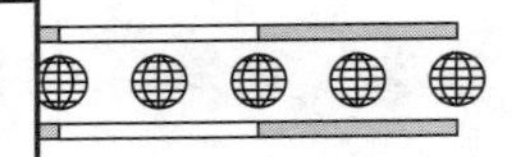

Some maps have imaginary lines that circle the Earth from east to west. These are called **lines of latitude**. Each line of latitude is named with a number of **degrees**. The **Equator** is 0°. All the other lines of latitude are north or south of the Equator. They go up to 90° N (north) or down to 90° S (south).

There are also imaginary lines that circle the Earth from north to south. These are called **lines of longitude**. Each line of longitude is named with a number of degrees, too. The **Prime Meridian** is 0°. All the other lines of longitude are east or west of the Prime Meridian. They go up to about 179° E (east) and 179° W (west). There is only one 180° line of longitude. Like the Prime Meridian and the Equator, it does not have a direction.

Together, the lines of latitude and longitude form a grid. By using lines of latitude and longitude, you can find any place on the Earth. Sometimes, a place is located between the lines. Then, you have to estimate its location.

Look at the map of South Dakota. Rapid City and Philip are both near 44° N latitude. Of course, they are not in the same place. By adding the longitude of each place, you can show exactly where it is. The latitude of a place is written first. The longitude is written next to it. Look at Watertown. We can say that Watertown is near 45° N, 97° W.

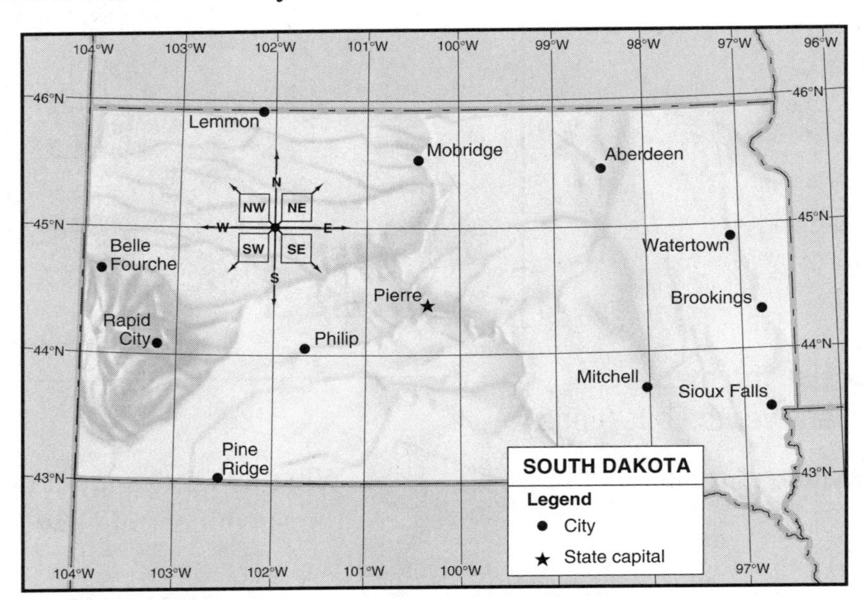

Directions Use the map to answer the questions.

1. Which city in South Dakota is located closest to 43° N?

____

2. Which city in South Dakota is located near 44° N, 98° W?

____

3. What are the latitude and longitude of Lemmon?

____

4. What is the state capital of South Dakota? What are its latitude and longitude?

____

Name ______________________ Date ______________

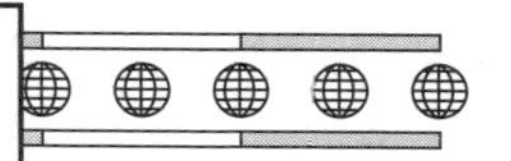

You're in Europe

The Prime Meridian runs through the site of the Royal Observatory at Greenwich, England, which is near London. Which other European countries does the Prime Meridian run through?

WESTERN EUROPE

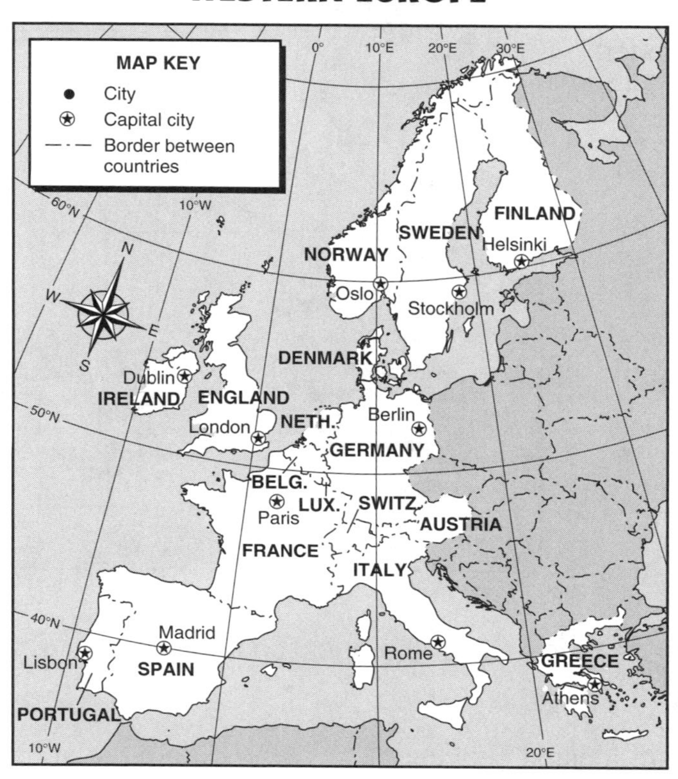

Directions Use the map to answer the questions.

1. What is the national capital of Greece?

____

2. What is the national capital of Sweden?

____

3. You travel from Germany to Spain. Which direction do you go?

____

4. What city is located at 60° N, 25° E?

____

5. Which city is located on the Prime Meridian?

____

6. What are the latitude and longitude of Oslo, Norway?

____

7. What are the latitude and longitude of Madrid, Spain?

____

Name ______________________ Date ______________

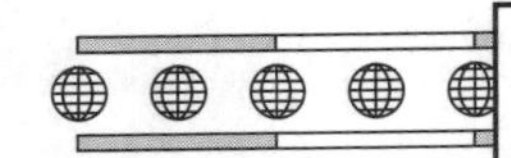

Where Are You Now?

You have learned about lines of latitude and longitude. Now let's use what you have learned. Look at the map of Nigeria. The numbers for the lines of latitude are at the left and right. The numbers for the lines of longitude are at the top and bottom. To help you find a place, move your finger along the lines. Remember, if a place is located between the lines, you have to estimate its location.

NIGERIA

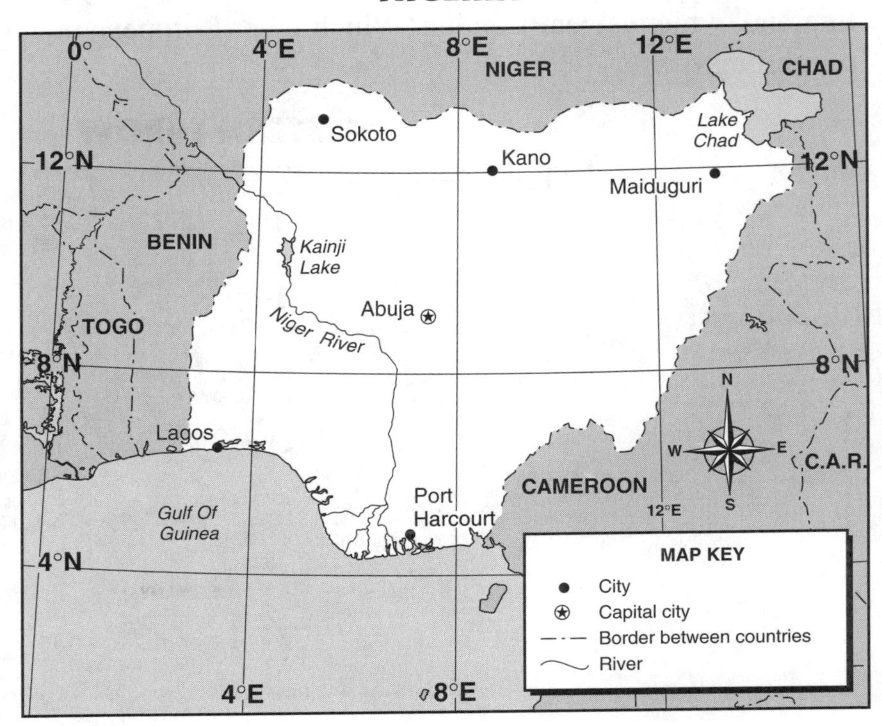

Directions Use the map to answer the questions.

1. Which three cities are located between 4° E and 8° E?

____

____

____

2. Which city has a latitude and longitude of about 6° N, 3° E?

____

3. What are the approximate latitude and longitude of Abuja?

____

4. What are the approximate latitude and longitude of Sokoto?

____

Activity

Find where you live on the map of the United States. If you cannot find your town, find a city near it. Write down the lines of latitude and longitude near where you live. Remember to write latitude first. Don't forget to include the directions of *N, S, E,* or *W.* Then, go to this address on the Internet:

http://www.fourmilab.ch/cgi-bin/uncgi/Earth

This is the web address for EarthCam. At this site you can type in your latitude and longitude. Then, you can see your area from a camera in space!

Name ______________________________ Date ______________

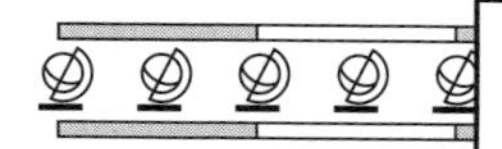

Globes

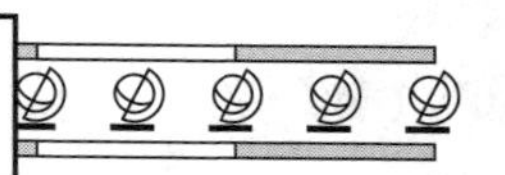

The Earth is a sphere, shaped like a round ball. A **globe** is a model of the Earth, so a globe is shaped like a round ball, too. A globe shows the large, main pieces of land, called **continents**. There are seven continents. A globe shows the main bodies of water, called **oceans**. There are four oceans. A globe also shows many of the seas, the smaller bodies of water.

Globes have a compass rose to help you to find directions. Some globes also have a map key and distance scale. Most globes also have lines of latitude and longitude.

Do you have a globe? If you do, look at it. If not, look at the map below. At the top of the globe is the North Pole. At the bottom of the globe is the South Pole. Find the Prime Meridian and Equator on the globe.

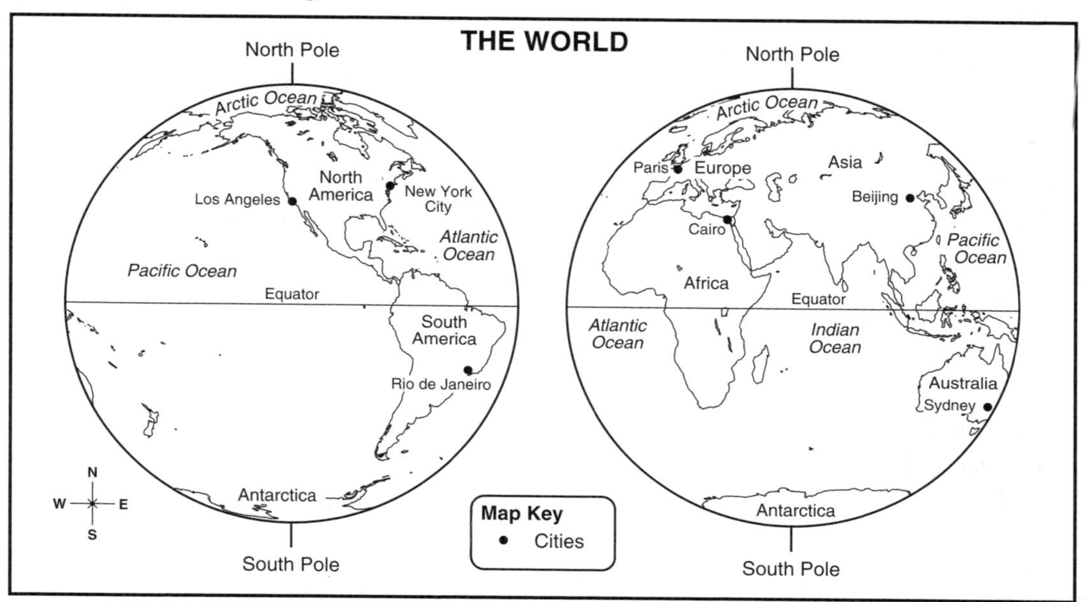

Directions Study a globe or the map. Use it to answer the questions.

1. What are the names of the seven continents?

______________________ ______________________

______________________ ______________________

______________________ ______________________

____

2. What are the names of the four oceans?

______________________ ______________________

______________________ ______________________

3. On which continent is Beijing located?

____

4. On which continent do you live?

____

Name ______________________ Date ______________

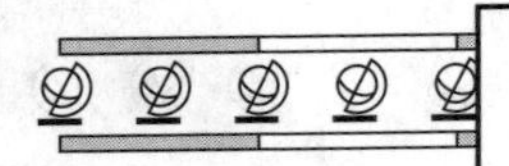

World Maps

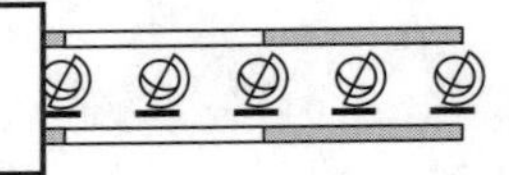

Unlike globes, world maps are flat. On a world map, South America is just shown as being east of Australia. On a globe, you can see that Australia is really on the opposite side of the world from South America. Globes are good for showing where places really are on the Earth. World maps are good for helping you to know what all those places are.

Look at the world map on page 6. You can see all the continents and oceans. On a globe, you would see only half of those things. You would have to turn the globe to see the rest.

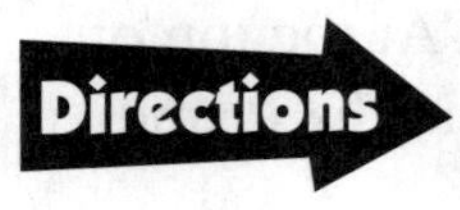

Study the world map on page 6. Use the map to answer the questions. Darken the circle by your answer choice.

1. Which continent is located below 60° S?
Ⓐ South America
Ⓑ Australia
Ⓒ Africa
Ⓓ Antarctica

2. On which continent is the United States located?
Ⓐ Europe
Ⓑ North America
Ⓒ South America
Ⓓ Africa

3. Which continent is an island?
Ⓐ Europe
Ⓑ Asia
Ⓒ South America
Ⓓ Australia

4. To travel directly from South America to Europe, which direction must you go?
Ⓐ northwest
Ⓑ southeast
Ⓒ northeast
Ⓓ southwest

5. Through which of these continents does the Prime Meridian run?
Ⓐ Europe and Asia
Ⓑ Asia and Africa
Ⓒ North America and South America
Ⓓ Europe and Antarctica

6. To travel directly from Australia to Africa, which ocean must you cross?
Ⓐ Arctic Ocean
Ⓑ Atlantic Ocean
Ⓒ Pacific Ocean
Ⓓ Indian Ocean

7. Which ocean touches both South America and Asia?
Ⓐ Arctic Ocean
Ⓑ Atlantic Ocean
Ⓒ Pacific Ocean
Ⓓ Indian Ocean

8. Which ocean is the most northern?
Ⓐ Arctic Ocean
Ⓑ Atlantic Ocean
Ⓒ Pacific Ocean
Ⓓ Indian Ocean

Put a big dot on the world map to show where you live.

Name ______________________ Date ______________

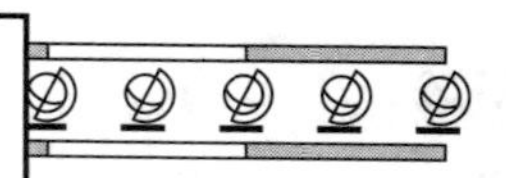

Hemispheres

A **hemisphere** is a half of a globe. The Earth has four hemispheres. This sounds like the Earth has four halves. It does, in a way. The Equator splits the Earth into two halves, the Northern Hemisphere and the Southern Hemisphere. The Prime Meridian splits the Earth into two hemispheres, too. Those two are the Western Hemisphere and the Eastern Hemisphere. So the Earth does have four halves! Most places are in two hemispheres. North America is in the Northern Hemisphere. It is also in the Western Hemisphere. Africa has places in all four hemispheres!

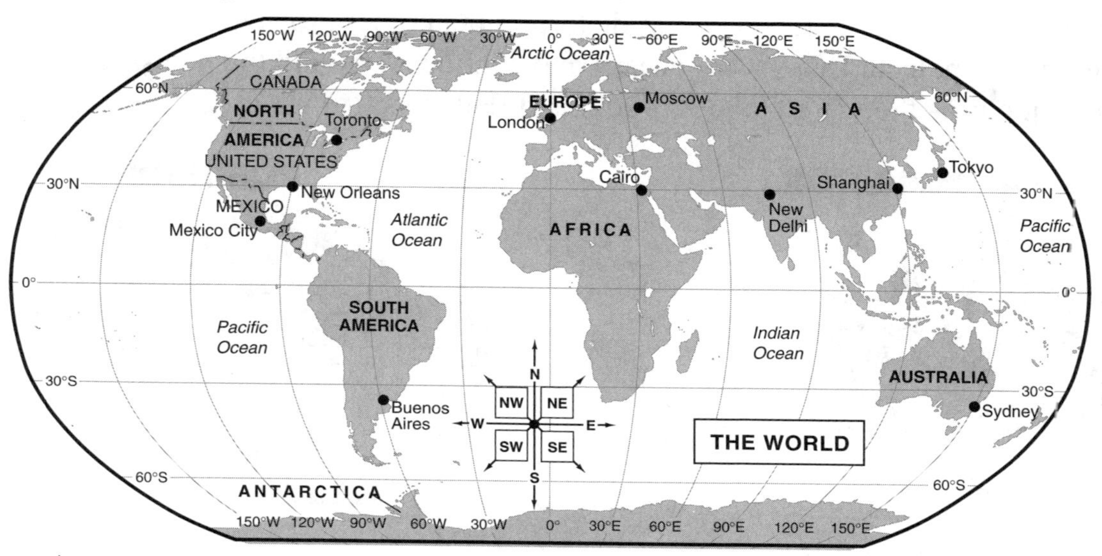

Directions Study the map. Use the map to answer the questions.

1. In which two hemispheres is Australia located?

____

____

2. In which three hemispheres is Europe located?

____

____

____

3. In which three hemispheres is South America located?

____

____

____

4. In which two hemispheres is the city of Moscow located?

____

____

5. In which two hemispheres is Mexico City located?

____

____

6. In which three hemispheres is Antarctica located?

____

____

____

Name ______________________________ Date ________________

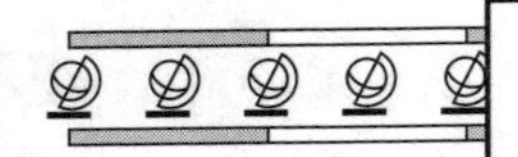

Map Projections

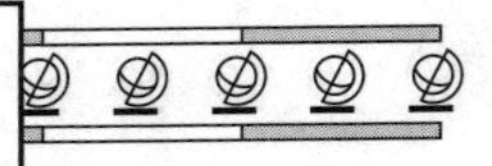

The Earth is a sphere, so the most accurate model is a globe. Most maps, though, are flat. Mapmakers, or **cartographers**, have tried many ways to make flat maps of the spherical Earth. Most of these ways cause distortions in the way the land masses really look. A **projection** is a way of showing Earth's curved surface on a flat map.

Look at the map projections below. You are probably most familiar with the Mercator projection. It shows the compass directions between places accurately. But distance and size on this projection are distorted, especially near the poles. The Robinson projection most closely resembles a globe. It shows less distortion than other projections. A polar projection is centered on a pole and shows the oceans and continents around the poles. An interrupted projection looks as though a globe was split and flattened.

MERCATOR

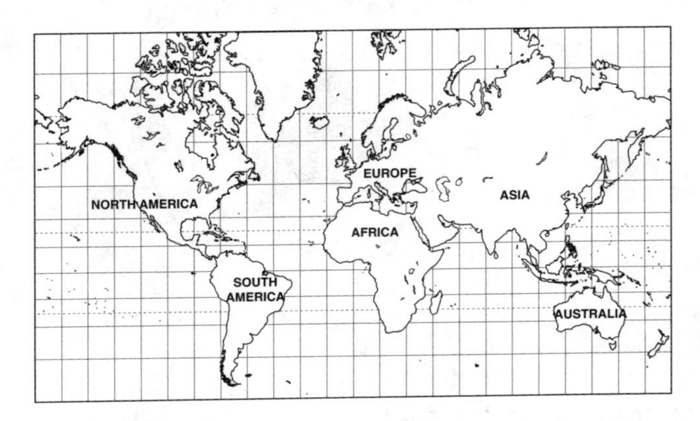

ROBINSON

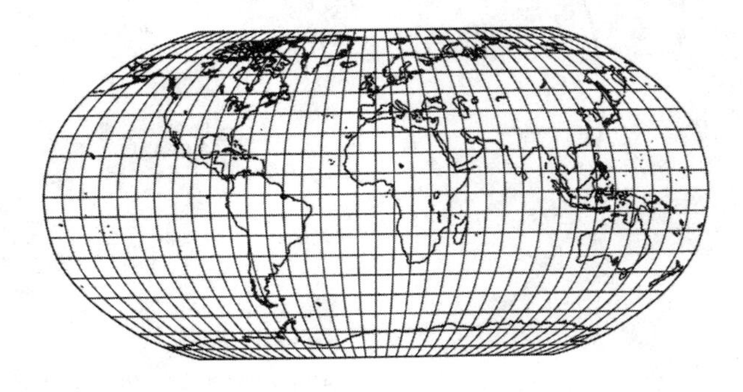

POLAR

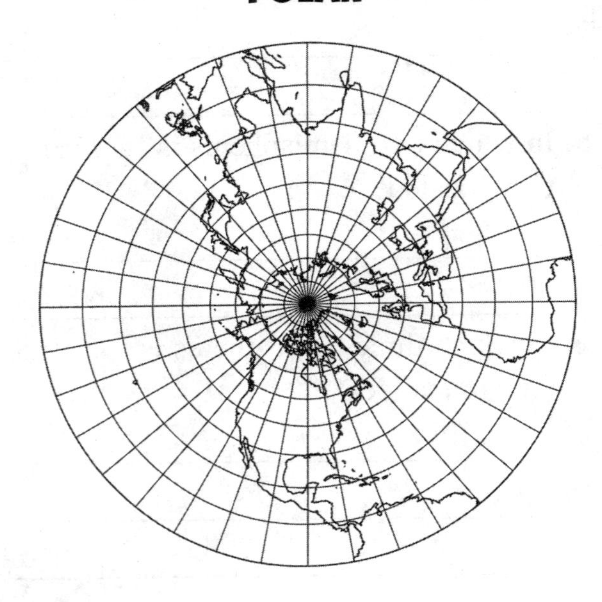

INTERRUPTED

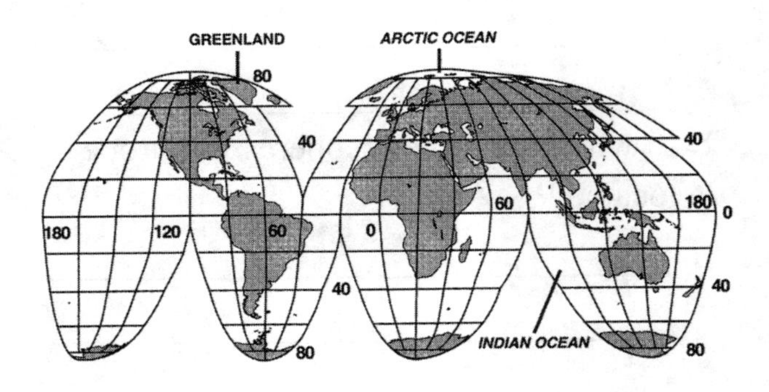

Go on to the next page.

Name ______________________________ Date ______________

Map Projections, page 2

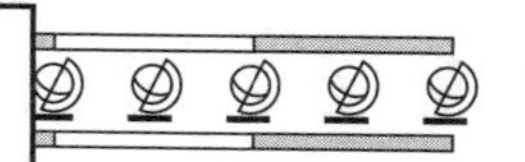

Study the map projections on page 36. Then, use the map projections to answer the questions.

1. Which projection looks like most world maps you have seen?

____

2. Which projection looks most like a globe?

____

3. Which projection looks as if it were drawn looking down on the North Pole?

____

4. Which projection looks like a cut-up globe?

____

5. Which projection can show only one hemisphere?

____

6. On which three projections does the Equator pass through the center of the map?

____

7. On which projection does the Equator form the circumference of the map?

____

8. Which projection do you think gives you the best understanding of the Earth's surface? Why?

____

____

____

____

Pretend that you are a cartographer. Develop your own projection method to draw a map of the world. Share your map with the class when you are finished.

Name ______________________ Date ____________

Political Maps

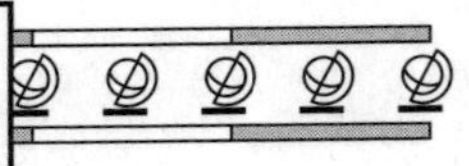

Political maps show the boundaries between political areas, such as states or nations. The map below shows the political boundaries for countries in the Middle East. Notice the symbol for a national capital in the map key.

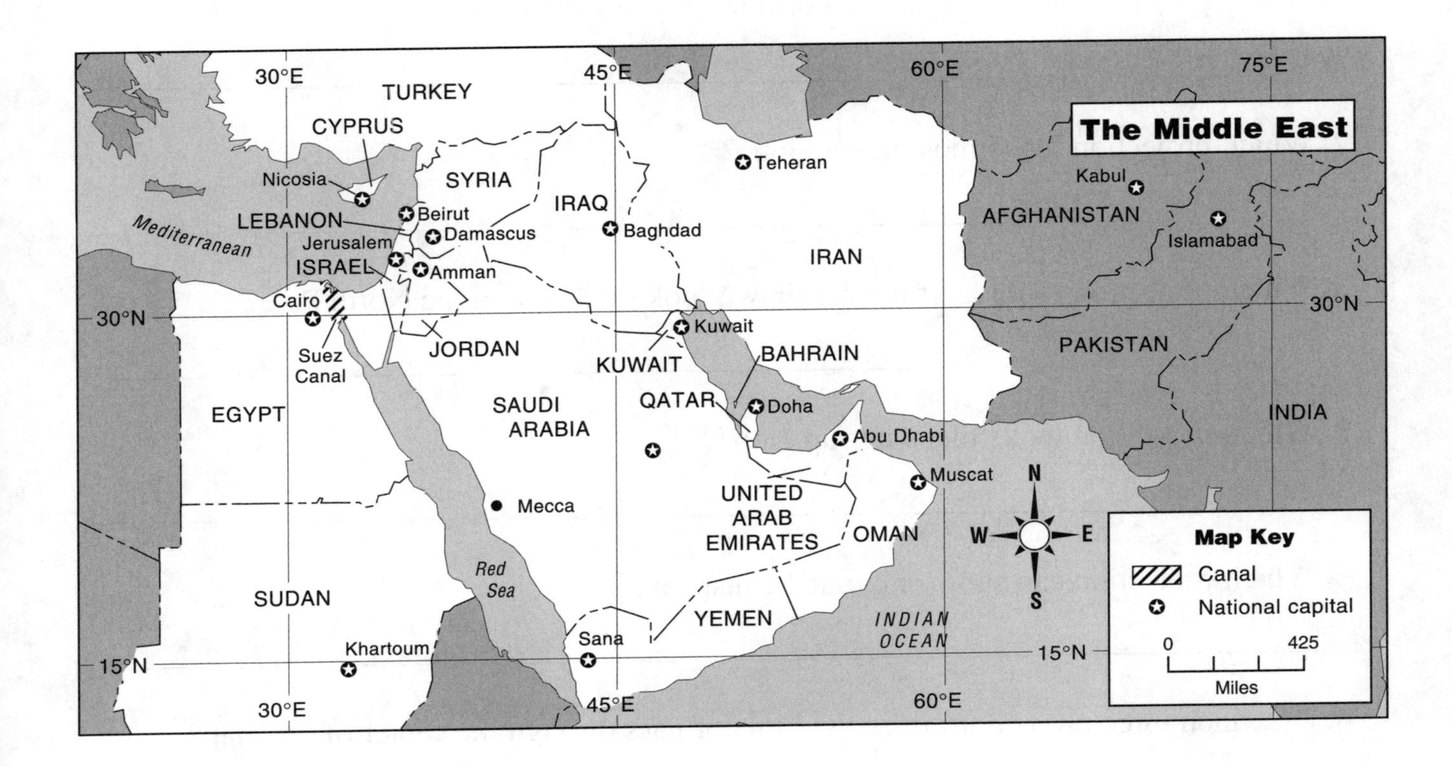

Directions Use the map to answer the questions.

1. Find the symbol of the national capital of Saudi Arabia. Label it Riyadh.

2. What is the national capital of Oman?

____

3. This map shows lines of latitude and longitude every ____________ degrees.

4. Which city is located at about 30° N, 32° E? ______________________

5. Which city is located at about 15° N, 33° E? ______________________

6. What are the latitude and longitude of Baghdad? ______________________

7. What are the latitude and longitude of Damascus? ______________________

8. In which country is the Suez Canal located? ______________________

Name ______________________ Date ______________

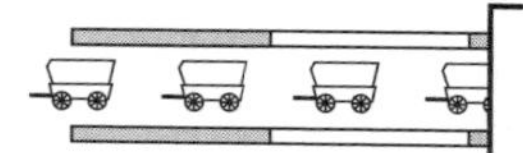

Historical Maps

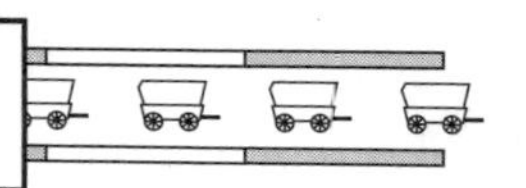

The world has changed over time. The names and boundaries of some states and countries have changed. Routes have changed, too. Some maps tell about places in the past. These maps are called **historical maps.** They help us to learn how the world used to be.

Look at the map below. It shows North America and the American Indian groups living there around 1600. The map shows the ways each group got its food and used the land. The map key uses a different kind of pattern for each food source.

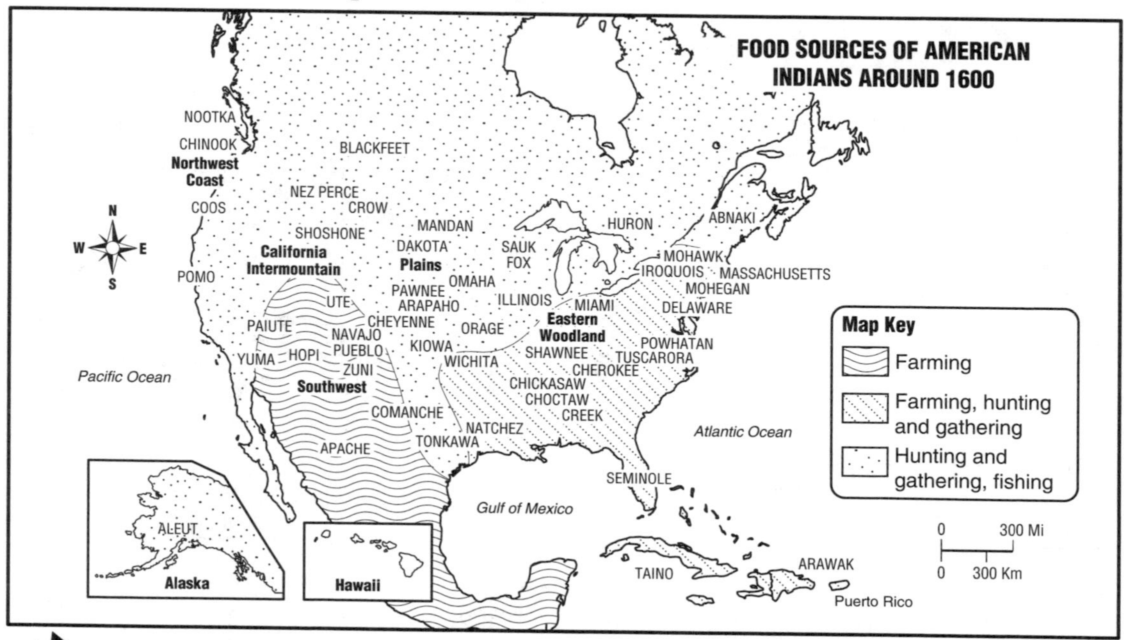

Directions Use the map to answer the questions.

1. Name two groups of American Indians in the northwest that might have used fishing as a food source.

____

____

2. In what part of North America did people rely on farming, hunting, and gathering for food?

____

3. In what part of North America did people rely only on farming for food?

____

4. How did the Shawnee people get their food?

____

5. How did most people living in North America get their food?

____

Activity

What do you think life would be like in 1600? Write a story telling about your struggles to find food for a day.

Name ______________________ Date ____________

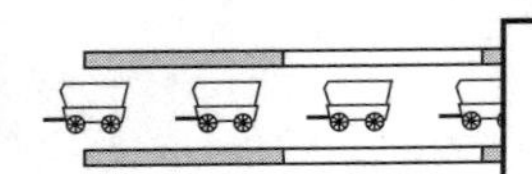

Looking Back

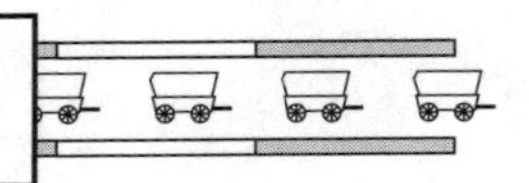

Historical maps can give us more information about events in the past. The map below shows some of the battles in the Revolutionary War. Study the map key and map index.

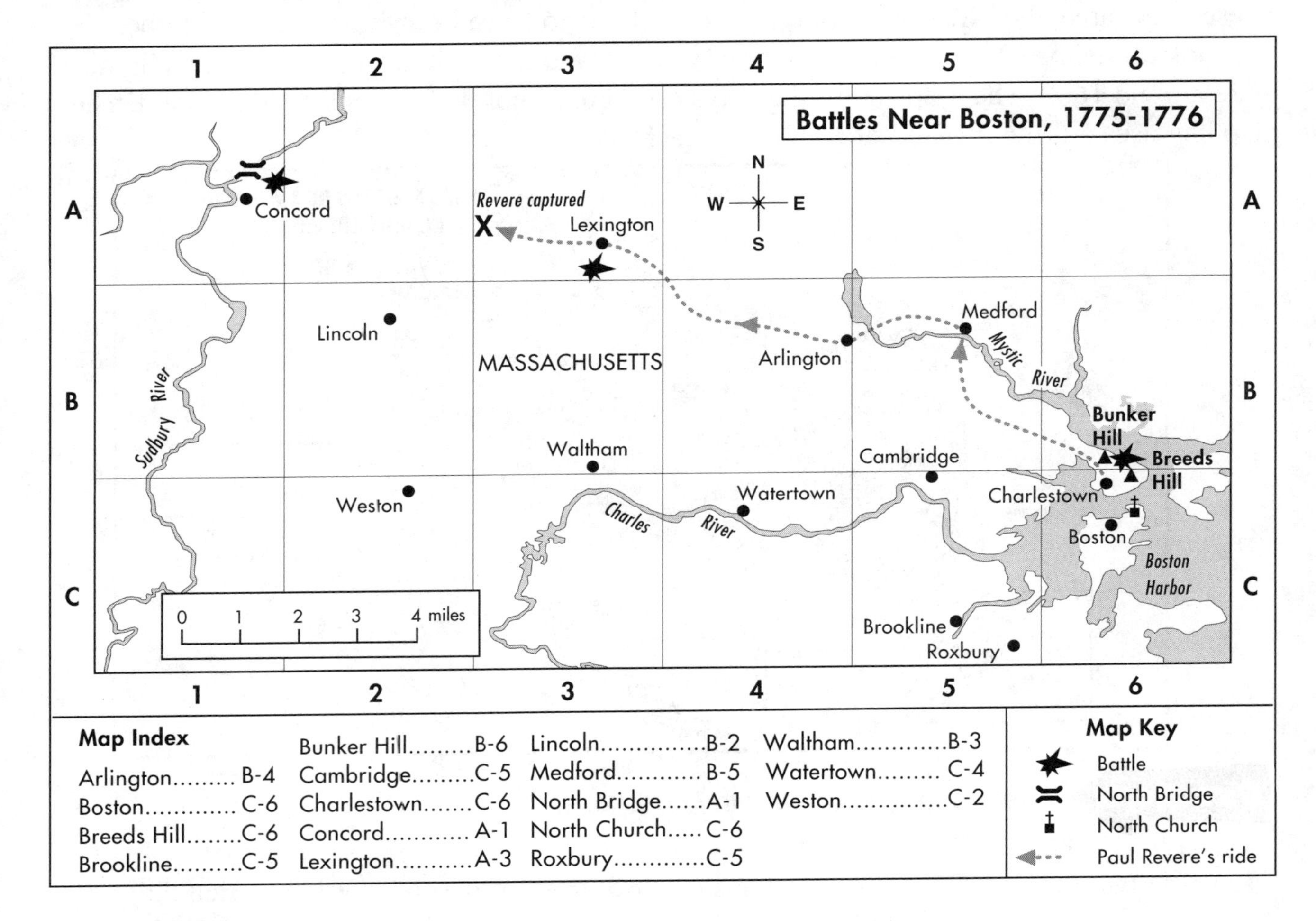

Directions Use the map to answer the questions.

1. How many battles are shown on the map? ______________________

2. In which grid square is Concord located? ______________________

3. In which grid square is Boston located? ______________________

4. In which town did Paul Revere begin his ride? ______________________

5. About how far from Lexington was Paul Revere when he was captured? ____________

6. About how far did Paul Revere ride altogether? ______________________

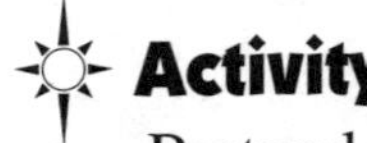

Activity

Pretend that you are Paul Revere. Write a story about your famous ride.

Name ______________________________ Date ______________

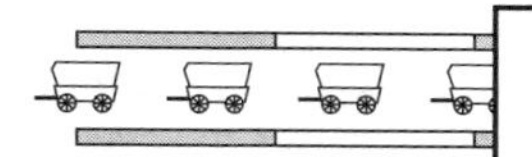

Past and Present

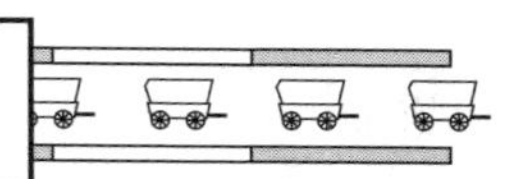

Historical maps are helpful to see how places and routes have changed over time. Look at the two maps below. The first map shows the Mississippi River valley during the Civil War. The second map shows the state of Mississippi in more recent times. You can compare the two maps to see how things have changed.

UNION ARMY IN THE MISSISSIPPI VALLEY

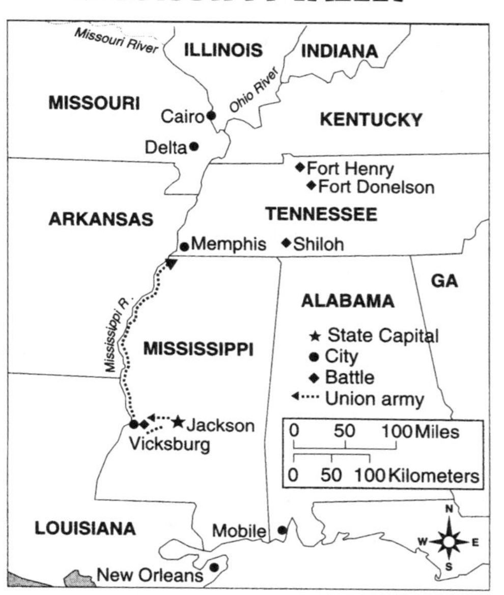

VICKSBURG AND SURROUNDING AREA

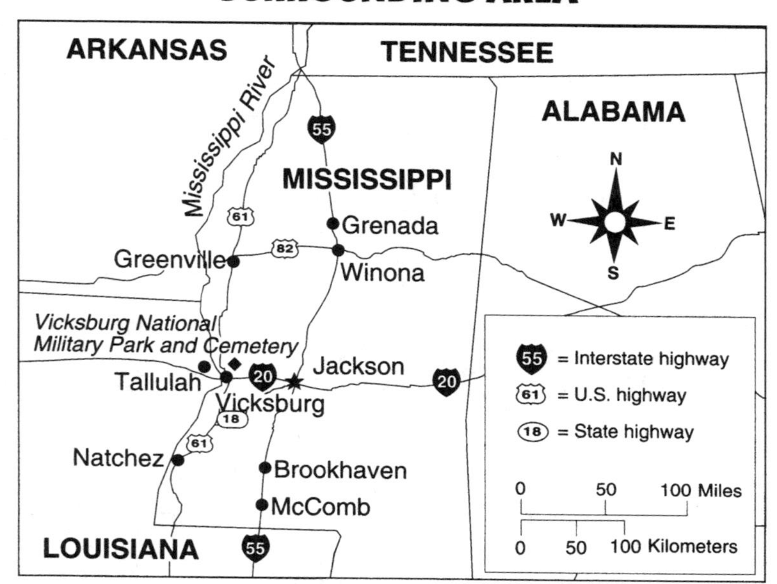

Directions Use the maps to answer the questions.

1. How many Civil War battles are shown on the left map?

____

2. What is the distance between Vicksburg and Jackson on the left map?

____

On the right map?

____

3. About how far did the Union Army have to march from Vicksburg to Memphis?

____

4. Through which cities did the Union Army probably pass on its march from Jackson to Memphis? (Hint: Use both maps to answer this question.)

____

____

____

5. Which map would you use if you wanted to visit Vicksburg today? Explain.

____

____

____

Name ______________________ Date ______________

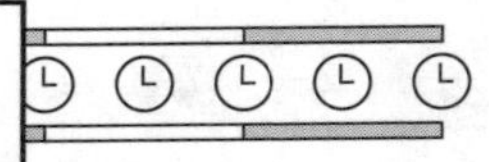

Time Zone Maps

The Earth is divided into 24 **time zones**. Each time zone is about 15 degrees wide. All the people in the same time zone set their clocks to the same time. Imagine that the time zones are numbered 1 to 24. The number 1 zone is at the International Date Line, or 180° longitude. As the number of the time zone increases by 1, the time becomes an hour later. The different time zones allow people around the world to experience a common event at about the same clock time. For example, most everyone in the world has sunrise at about 6:00 in the morning.

The United States has six time zones. As you move east from one time zone to another, the time gets an hour later.

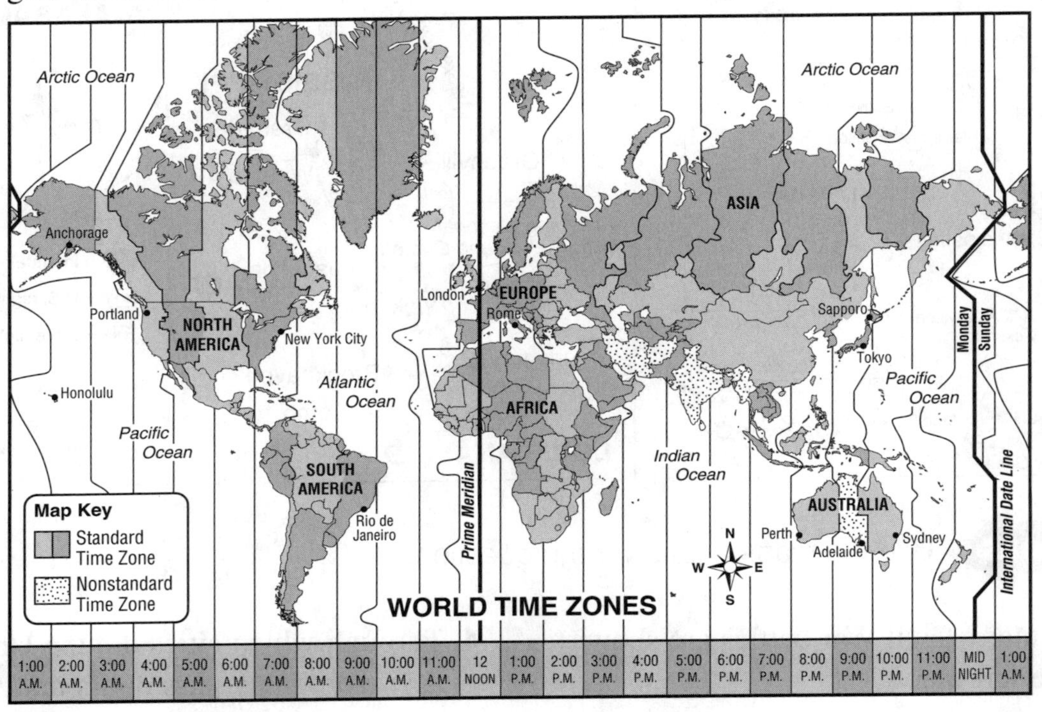

Directions Study the map. Use the map to answer the questions.

1. Through which oceans does the International Date Line pass?

____

2. When it is 4:00 A.M. in Portland, what time is it in Sydney?

____

3. When it is 9:00 P.M. in Tokyo, what time is it in Rio de Janeiro?

____

4. What does this difference in time suggest about the location of the two cities?

____

____

Activity

Write a story about going backward or forward in time. To what time period would you go? Include a drawing of your time machine.

Name ______________________________ Date ______________

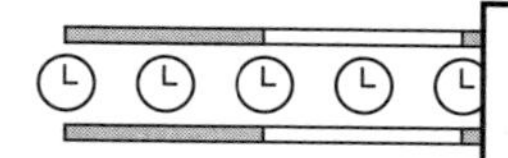

Cultural Maps

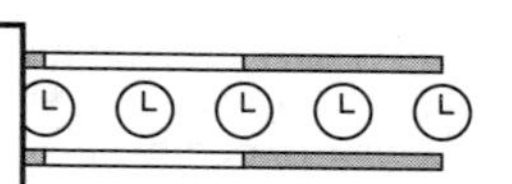

A **cultural map** tells about groups of people that live in a certain area. The map may tell about their lifestyle or ethnic background. It may point out the religion they practice or the language they speak. Look at the map below. It shows information about Hindus living in India in 1951. The map uses different kinds of shading to show the different percentages of Hindus living in an area.

HINDUS IN INDIA, 1951

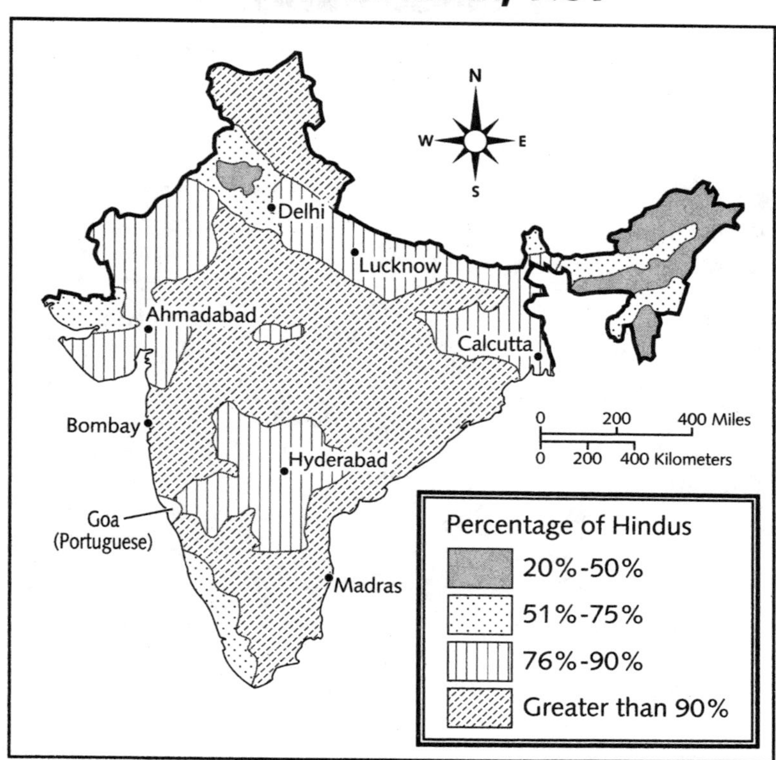

Directions Study the map. Use the map to answer the questions.

1. What percentage of the population was Hindu in the area around Madras?

____

2. What percentage of the population was Hindu in the area around Delhi?

____

3. What percentage of the population was Hindu in the area around Hyderabad?

____

4. Did any parts of India have a population of less than 20% Hindu? How can you tell?

____

____

____

5. Based on this map, what can you guess about the culture of India in 1951?

____

____

Name ______________________________ Date ______________

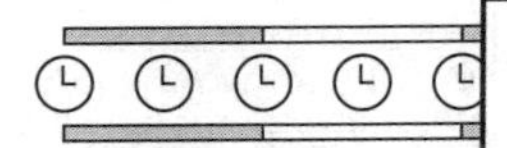

Vegetation Maps

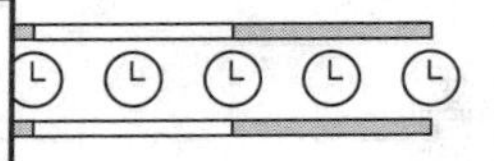

A **vegetation map** shows what kinds of vegetation, or plants, grow in a certain area. If you wanted to visit a place with a certain kind of plant growth, a vegetation map would help you to pick the place. The map key shows you how to identify each kind of vegetation. For example, black coloring on the map below stands for mixed forests.

VEGETATION MAP OF SOUTH AND SOUTHEAST ASIA

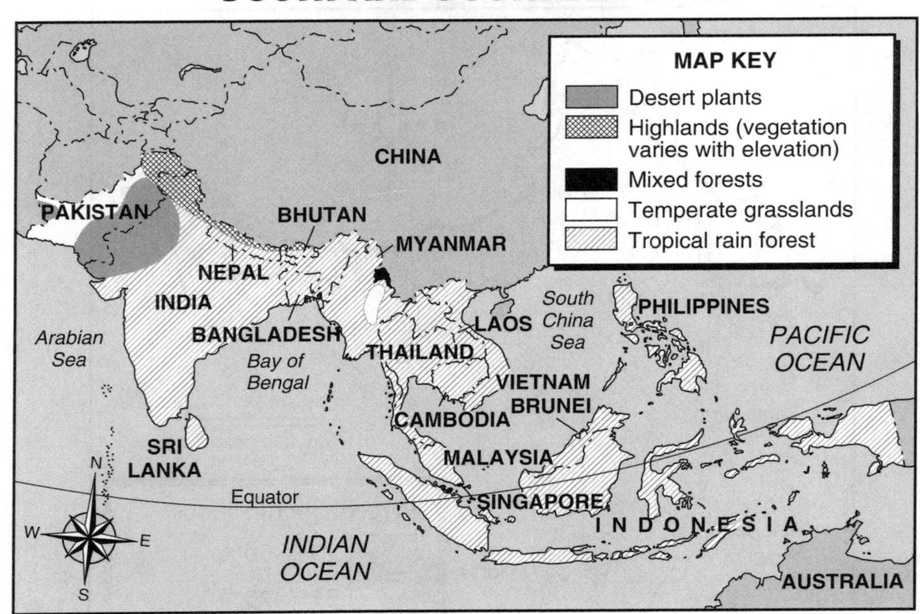

Directions Use the map to answer the questions.

1. What kind of vegetation does most of South and Southeast Asia have?

____

2. In which two countries can temperate grasslands be found?

____

____

3. Which country has mixed forests?

____

4. In what four countries is highland vegetation found?

____

____

____

____

5. What kind of vegetation is found along the boundary between Pakistan and India?

____

Activity

Use your neighborhood map from page 12. Add vegetation to it, such as trees and shrubs. Add the symbols for this vegetation to your map key.

Name ______________________ Date ______________

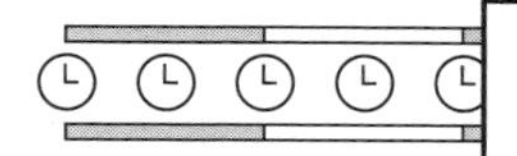

Climate Maps

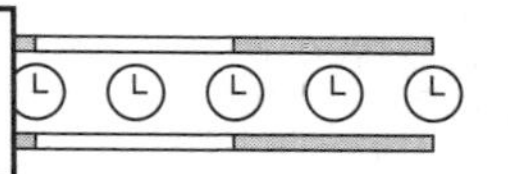

A **climate map** tells about the kind of climate a certain area has. Climate is the kind of weather a place has over a long period of time. The climate map shows which places are rainy and which places are dry. It also tells which places are hot and which are cooler. Study the map below. The map key uses different kinds of shading to show the different climate types.

CLIMATE MAP OF TURKEY

Directions Use the map to answer the questions.

1. What is the national capital of Turkey?

____

2. What kind of climate does the land along Turkey's two coasts have?

____

3. What kind of climate is common around Mt. Ararat?

____

4. What kind of winters does Ankara have?

____

5. What kind of summers does Izmir have?

____

6. What kind of climate do the mountainous regions have?

____

Name ______________________________ Date ______________

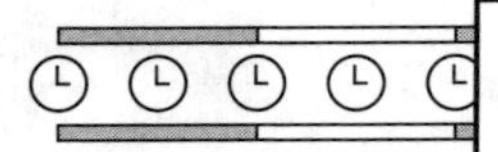

Comparing Maps

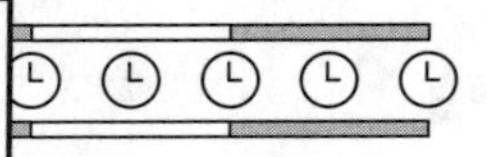

A **population map** shows how many people live in a place. This kind of map shows the density of population. Density of population means how many people live in a certain area, usually a square mile. The map on the left below shows the population of Brazil. Notice the map key. It shows the density scale, or how many persons per square mile.

A climate map tells about the kind of weather a certain area has. The map on the right below shows the climate of Brazil. Three climate types are listed in the map key. Each climate type is indicated by a pattern. Find each pattern on the map.

You can compare the two maps to learn things about Brazil. For example, you can find which type of climate people in Brazil prefer. Some of the largest cities in Brazil are located where the climate is mild. Find that area on the two maps.

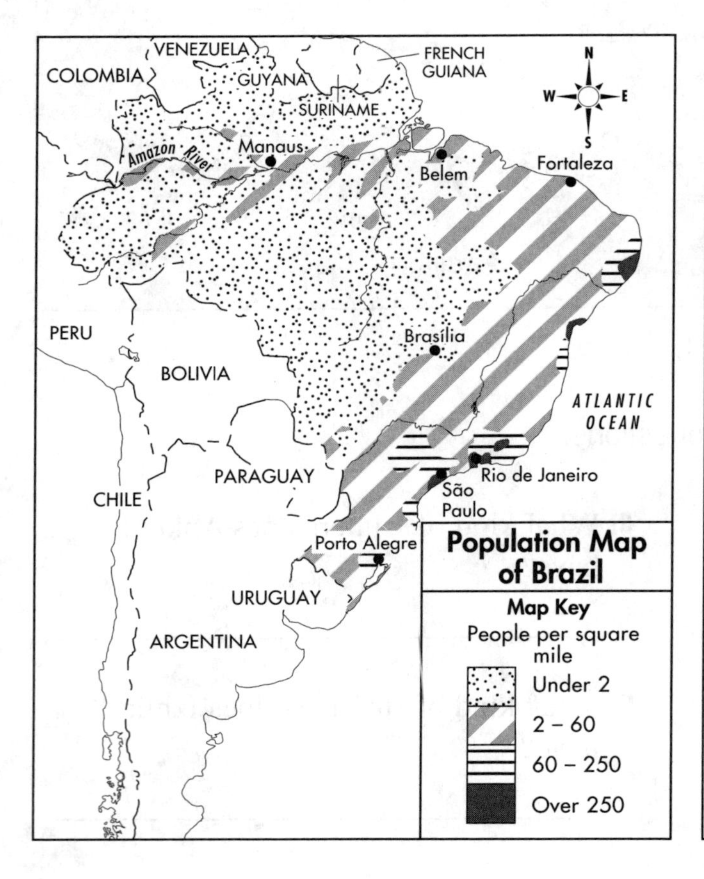

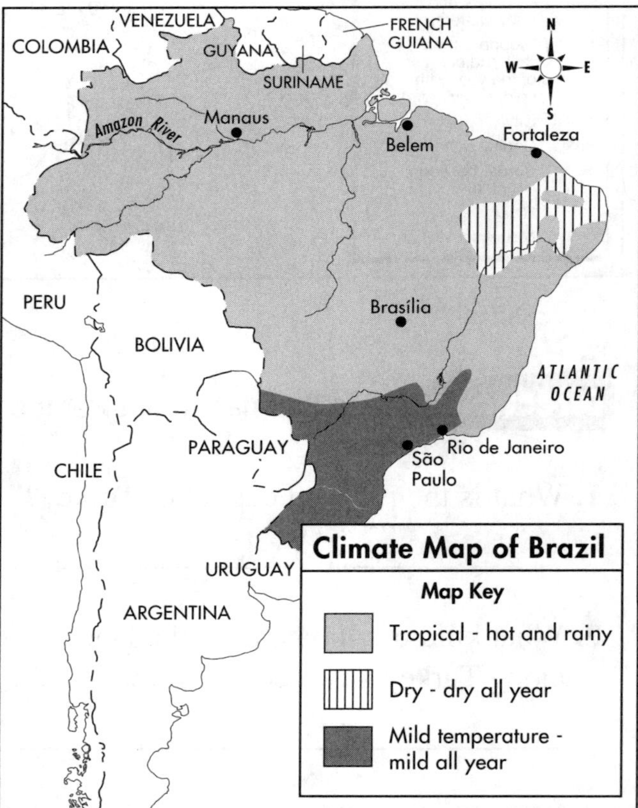

Name ______________________ Date ______________

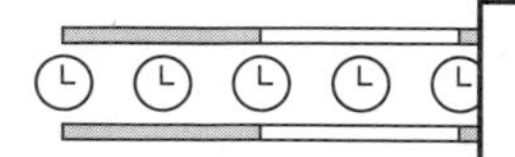

Comparing Maps, page 2

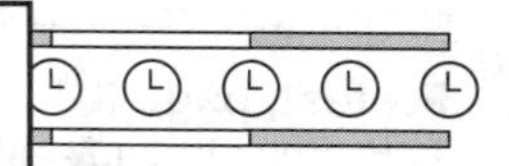

Directions Study the maps on page 46. Use the maps to answer the questions.

1. About how many people per square mile live in the area around Belem?

____

2. What kind of climate is found around Belem?

____

3. About how many people per square mile live in the area around Rio de Janeiro?

____

4. What kind of climate is found around Rio de Janeiro?

____

5. What kind of climate area does the Amazon River pass through?

____

6. Much of Brazil is covered in rain forest. What kind of climate would probably be found in the rain forest?

____

7. Do more people live near the coast or in the middle of Brazil?

____

8. Why do you think many people in Brazil choose to live near the coast?

____

____

____

____

Activity

Would you like to live in the rain forest? What do you think life would be like there? On another piece of paper, write a story about daily life in the rain forest. Include a picture with your story.

Answer Key for Map Skills, Grade 6

Pretest, page 3
1. Pacific Ocean, **2.** Mexico, **3.** Houston, **4.** Memphis, **5.** Denver, Pittsburgh, Philadelphia, **6.** Miami, **7.** Minneapolis
Posttest, page 4
1. C, **2.** B, **3.** D, **4.** C, **5.** D, **6.** A
page 7
1. east, **2.** north, **3.** north and south, **4.** east and west
page 9
1. A, **2.** D, **3.** D, **4.** C
page 10
1. Old Town, **2.** school, **3.** bike path, **4.** northeast
page 11
1. B-4, **2.** C-5, **3.** Atlantic Ocean, **4.** B-3, **5.** Cuba
page 12
Map Index: Acme Supermarket: B-2; City Hall: A-1; Fire Station: C-2; Jefferson High School: A-3; Mini-mall: C-1; Mirror Lake: D-2; Police Station: A-1; Post Office: B-1; Washington Elementary School: B-3
page 13
1. Sacramento, **2.** Canada, **3.** northwest, **4.** southwest, **5.** Juneau, **6.** Honolulu, **7.** Salem
page 14
1. United States Map, **2.** Washington, D.C., **3.** Concord, **4.** Santa Fe, **5.** Canada, **6.** Mexico, **7.** southeast, **8.** southwest, **9.** Boise, Idaho, **10.** Tallahassee, Florida
page 15
1. $1\frac{1}{2}$, **2.** 450, **3.** 720, **4.** $\frac{1}{2}$, **5.** 150, **6.** Check students' maps.
page 16
1. about 8 miles, **2.** about 15 kilometers, **3.** about 20 miles, **4.** about 30 miles, **5.** about 15 kilometers, **6.** about 10 miles, **7.** about 15 miles, **8.** about 5 kilometers
page 17
1. larger map, **2.** about 90 miles, **3.** smaller map, **4.** about 900 kilometers, **5.** larger map
page 18
1. State Highway 7, **2.** 30 miles, **3.** north on State Highway 50 to Interstate Highway 3, north on Interstate Highway 3 to Bullock, **4.** north and south, or northwest and southeast
page 19
1. Answers may vary: Go north on Washington Street to Decatur Avenue, then northwest on Decatur Avenue to Peachtree Street, then north on Peachtree Street to Peachtree Center., **2.** about a mile, **3.** Answers may vary: Go east on International Boulevard to Piedmont Avenue, then south on Piedmont Avenue to Edgewood Avenue, then east on Edgewood Avenue to the site., **4.** State Archives
page 20
1. B, **2.** A, **3.** C, **4.** D
page 21
1. 221, **2.** 240, **3.** 457, **4.** 581, **5.** 228, **6.** 224
page 22
1. about 3 hours, **2.** Answers may vary: Go east on Interstate Highway 90 to U.S. Highway 14, then east on U.S. Highway 14 to U.S. Highway 281, then north on U.S. Highway 281 to Aberdeen, **3.** about 8 hours, **4.** about 165 miles, **5.** about 3 hours, **6.** 2 P.M.
page 23
1. Sonoran Desert, **2.** Central Plains, **3.** Santa Fe, **4.** Ouachita Mountains, **5.** Coastal Plain, **6.** Sonoran Desert, Painted Desert, Chihuahuan Desert, **7.** Colorado Plateau, Edwards Plateau
page 24
1. west, **2.** southeast, **3.** higher, **4.** 2,000 to 5,000 feet, **5.** Coastal Plain, **6.** 0 to 250 feet
page 25
1. Answers may vary: The left map gives information on landforms in Italy, and the right map gives information on major roads in that country., **2.** Answers may vary: The mountains are in the center of the country, and the cities are on both sides, at the foot of the mountains., **3.** Answers may vary: The mountains are in the center of the country, and the roads are on both sides, at the foot of the mountains, connecting the cities.
page 26
1. A, **2.** C, **3.** B, **4.** C
page 27
1. C, **2.** A, **3.** B, **4.** C
page 28
1. B, **2.** A, **3.** D, **4.** C
page 29
1. northeast, **2.** Hammonton, **3.** truck farming, **4.** cranberries, blueberries, **5.** Vineland, because the land is used more for farming there.
page 30
1. Pine Ridge, **2.** Mitchell, **3.** about 46° N, 102° W, **4.** Pierre; about 44° N, 100° W
page 31
1. Athens, **2.** Stockholm, **3.** southwest, **4.** Helsinki, Finland, **5.** London, England, **6.** 60° N, 10° E, **7.** 40° N, 5° W
page 32
1. Sokoto, Abuja, Port Harcourt, **2.** Lagos, **3.** 9° N, 7° E, **4.** 13° N, 5° E
page 33
1. Asia, Africa, Australia, Antarctica, Europe, North America, South America, **2.** Arctic Ocean, Atlantic Ocean, Indian Ocean, Pacific Ocean, **3.** Asia, **4.** Answers may vary: North America.
page 34
1. D, **2.** B, **3.** D, **4.** C, **5.** D, **6.** D, **7.** C, **8.** A
page 35
1. Southern Hemisphere, Eastern Hemisphere, **2.** Northern Hemisphere, Eastern Hemisphere, Western Hemisphere, **3.** Northern Hemisphere, Southern Hemisphere, Western Hemisphere, **4.** Northern Hemisphere, Eastern Hemisphere, **5.** Northern Hemisphere, Western Hemisphere, **6.** Southern Hemisphere, Eastern Hemisphere, Western Hemisphere
page 37
1. Mercator, **2.** Robinson or Polar, **3.** Polar, **4.** Interrupted, **5.** Polar, **6.** Mercator, Robinson, Interrupted, **7.** Polar, **8.** Answers will vary.
page 38
1. Check students' maps., **2.** Muscat, **3.** 15, **4.** Cairo, Egypt, **5.** Khartoum, Sudan, **6.** about 34° N, 45° E, **7.** about 34° N, 37° E, **8.** Egypt
page 39
1. Students should name two of these: Nootka, Chinook, Coos, Pomo., **2.** southeast, **3.** southwest, **4.** farming, hunting, and gathering, **5.** Answers may vary: hunting and gathering.
page 40
1. three, **2.** A-1, **3.** C-6, **4.** Charlestown, **5.** about 2 miles, **6.** about 12 miles
page 41
1. four, **2.** about 50 miles; about 45 miles, **3.** about 200 miles, **4.** Answers may vary: Vicksburg, Greenville., **5.** The right map, because it shows the area as it is today.
page 42
1. Pacific Ocean and Arctic Ocean, **2.** 10:00 P.M., **3.** 9:00 A.M., **4.** They are 180° apart, or on opposite sides of the Earth.
page 43
1. greater than 90%, **2.** Answers may vary: 51 to 75% or 76 to 90%., **3.** 76 to 90%, **4.** Yes; in the northeast part of the country and also northwest of Delhi; those areas have solid shading., **5.** Answers may vary: That the population is mostly Hindu.
page 44
1. tropical rain forest, **2.** Pakistan and Myanmar, **3.** Myanmar, **4.** Nepal, Bhutan, India, and Pakistan, **5.** desert plants
page 45
1. Ankara, **2.** Mediterranean climate, **3.** Steppe climate, **4.** cold and dry, **5.** hot and dry, **6.** Mediterranean climate
page 47
1. 2 to 60, **2.** tropical, **3.** Answers may vary: over 250., **4.** mild, **5.** tropical, **6.** tropical, **7.** coast, **8.** Answers will vary: Students might suggest that the closeness of the ocean allows for easier movement or more job opportunities, or that most of the major cities of Brazil are near the coast.